Why Does It Still Hurt?

Paul Biegler is a journalist, academic, and former doctor specialising in emergency medicine. His health and science writing has been published in *The Age*, *The Sydney Morning Herald*, *Good Weekend*, *The Australian Financial Review*, *Cosmos*, *New Philosopher*, and *Arena*, and he is the author of *The Ethical Treatment of Depression*, which won the Australian Museum Eureka Prize for Research in Ethics.

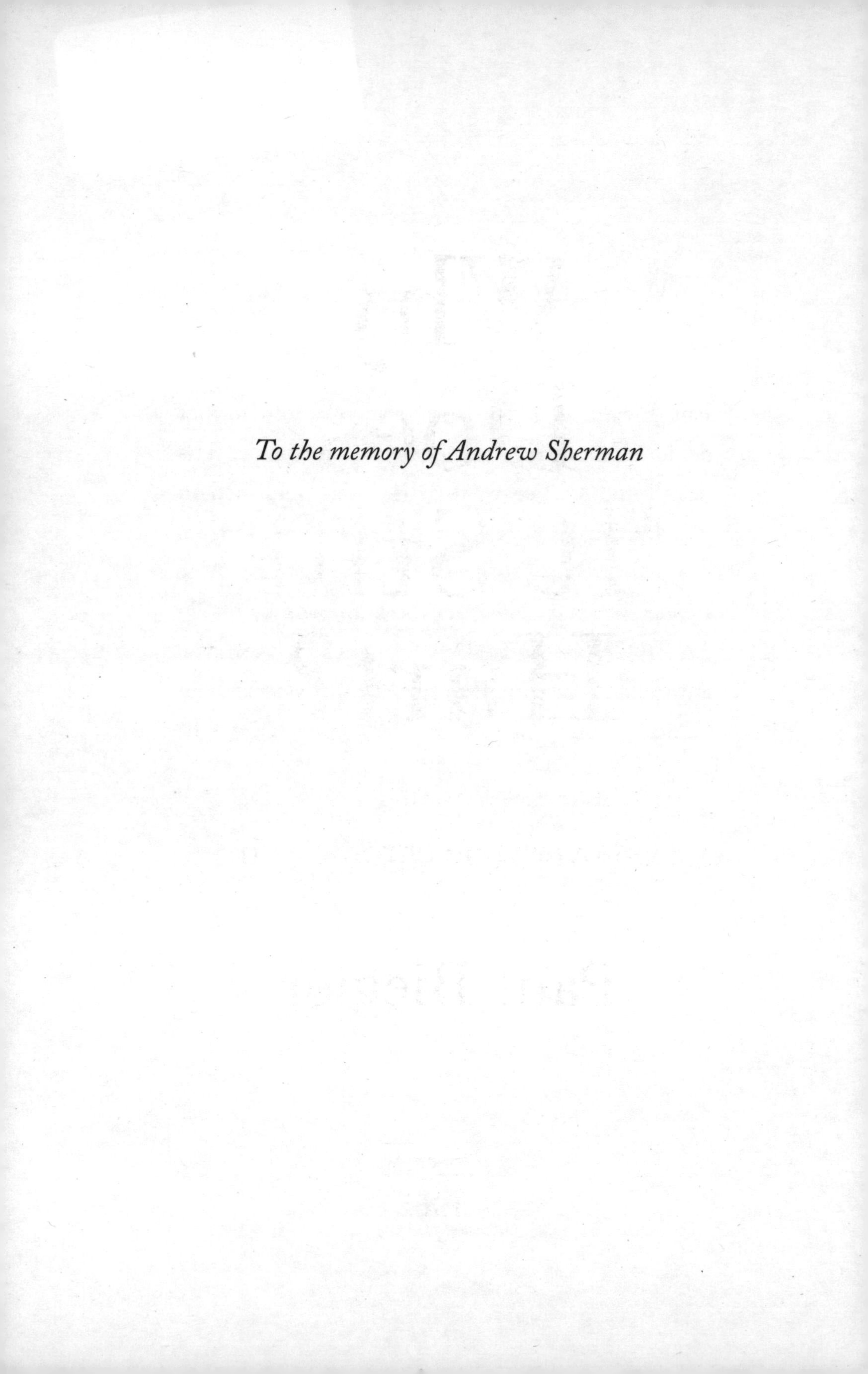

To the memory of Andrew Sherman

Why Does It Still Hurt?

how the power of knowledge can overcome chronic pain

Paul Biegler

SCRIBE
Melbourne • London

Scribe Publications
2 John St, Clerkenwell, London, WC1N 2ES, United Kingdom
18–20 Edward St, Brunswick, Victoria 3056, Australia
3754 Pleasant Ave, Suite 100, Minneapolis, Minnesota 55409, USA

Published by Scribe 2023

Copyright © Paul Biegler 2023

All rights reserved. Without limiting the rights under copyright reserved above, no part of this publication may be reproduced, stored in or introduced into a retrieval system, or transmitted, in any form or by any means (electronic, mechanical, photocopying, recording or otherwise) without the prior written permission of the publishers of this book.

The moral rights of the author have been asserted.

The advice in this book is not intended to replace the services of trained health professionals or be a substitute for medical advice. You are advised to consult with your health care professional with regard to matters relating to your health, and in particular regarding matters that may require diagnosis or medical attention.

Typeset in Adobe Caslon Pro 12.5 pt/19 pt by the publishers.

Printed and bound in the UK by CPI Group (UK) Ltd, Croydon CR0 4YY

Scribe Publications is committed to the sustainable use of natural resources and the use of paper products made responsibly from those resources.

978 1 914484 15 5 (UK edition)
978 1 957363 27 1 (US edition)
978 1 922585 23 3 (Australian edition)
978 1 922586 88 9 (ebook)

Catalogue records for this book are available from the National Library of Australia and the British Library.

scribepublications.co.uk
scribepublications.com
scribepublications.com.au

Contents

Introduction

The Card Trick

It was one of those searing summer days in Australia where you leave the protective dome of your car and know that your scalp is going to get fried on the ten-foot scuttle across the pavement to the nearest shade. My target, on this glaring Tuesday morning in November 2019, was the sheltered portico of an anonymous glass-fronted building in the Melbourne suburb of Moorabbin; that's an Aboriginal word meaning 'resting place', but, in these modern times, the place is a hive of activity, from the workaday hum of cheap Chinese restaurants to the whir of auto repair shops in a series of drab industrial estates. I beelined for the sliding glass doors and slipped into the cool confines of the building's atrium, where I did a quick recce and spotted the sign I was looking for: Rehabilitation Medicine Group. Inside, neutral walls were splashed with the contemporary hues of regulation office art; there was a reception counter and a row of sleek white chairs smudged with the just-discernible grey of repeat users. 'Dr Fried's 11 o'clock?' chimed the receptionist who, when I nodded assent, motioned me to sit down.

There is something about doctors' waiting rooms that, like the barber shop and airline travel, saps your will and inspires a kind of helplessness as you hand control to the physician, hairdresser, or pilot, as the case may be. I entered that fugue state, fidgeted with my phone, then lapsed into a meditation on why I was sitting here waiting to see a pain expert. And whether I really needed to be. I consider myself to be medically literate. I had been, after all, a doctor for 20 years, a decade of which I worked as an emergency physician, mopping up everything that disease and trauma throws at a person to put them at death's door. I've also conducted a decent slab of my own research, including a PhD and a postdoc on how our decision-making gets skewed by things as varied as depression and the subconscious tweaks of advertising. To top it off, I'm a science journalist, with the latest research filling my inbox on a daily basis. With a CV like that, you'd think I'd be in the box seat to nail the answer to my own medical dilemma. But there I sat, an ex-medico stuffed full of knowledge, in a state of hopeless confusion.

The source of my inner turmoil had begun on a wintry morning five months earlier. I'd woken with a sore right knee and had to push through tightness and a nagging ache when I walked my kids to school. I put it down to basketball and overdoing my regular jogs through the park. But five weeks in, I'd had enough. I happen to have a mate who's a physiotherapist and an expert diagnostician, so I decided it was time to get professionally reacquainted. Paolo is a tall, athletic man, whose dark hair and beard I've watched grey in recent years as we've both moved into our 50s. I managed to catch Paolo in his rooms in bayside Melbourne, where he greeted me with the relaxed demeanour of a friend, tinged with the subtle gravitas of the health professional. I had to climb a flight of stairs to get to his office and did it wincing, holding onto the handrail. Paolo put my knee through its paces, with some tests to see how far I could bend it and whether I could squat on that leg. It hurt. Enough to bring tears to my eyes. But pain, it seemed, was a prerequisite of the diagnosis; I had torn, Paolo said, a piece of cartilage in my knee called the medial meniscus. And he was right — the diagnosis was confirmed a few weeks later with an MRI scan. But making the diagnosis, it would turn out, was the easy part. The hard problem was what, exactly, to do about it?

I know an orthopaedic surgeon with impeccable credentials, a former colleague whom I hold in high regard, so I consulted him, and his advice was clear: surgery to trim the tear would help my pain, which would otherwise continue a waxing and waning course over an unspecified period of time. In fact, he'd had the very same surgery himself, with an excellent result. It was a gold-plated recommendation, and I signed the paper then and there to have the operation at a local public hospital. The lengthy machinations of the public health system, however, meant I was in for a wait, possibly months, and, while that time ever-so-slowly elapsed, a number of things happened. I gave up my beloved jogging and quit scratch basketball with my young son. I favoured my good leg more and more, and the quads on my injured side gradually weakened and wasted away. But I also hoovered up a small mountain of medical research about meniscal tears, which, disconcertingly, left me plagued with second thoughts. Was going under the knife, I wondered, really the best option?

So what did I learn that led me to question the surgeon's advice? It started with a better understanding of that little piece of me that was up for surgery. The meniscus is the shock-absorbing cartilage in the knee joint. The structure not only disperses the hefty forces that are driven through the knee when you walk, run, or jump, but also has the vital task of helping the bottom end of the thigh bone — the femur — slide over the top end of the lower leg bone — the tibia — when you bend your knee. There are two of these menisci, a 'medial' one on the inner side and a 'lateral' one on the outer side of each knee. If you enlarged them, they'd be a skateboarder's dream: they look like a half-pipe designed by Antoni Gaudí, with a wickedly high central lip that drops down to a more navigable height at front and back.

But if you ever have to take a job in the human body, don't be a meniscus. The forces that pass through the knee would test Atlas, ranging from one and a half times your body weight on a gentle walk, to three times your weight going up and down stairs. Go for a run and eight times your weight can pass through the knee. A good chunk of that compressive force is channelled through the meniscus and especially, when you're running, the back, or 'posterior', of the medial meniscus. Which is precisely where my tear was. Now, at this point, you'd think the treatment would be crystal clear. People have been doing this injury for ages, it has been thoroughly mapped by MRI scans for over three decades, and surgeons have been fixing it with a procedure called a partial meniscectomy for even longer. Yet the best treatment for a meniscal tear is as clear as mud, and the reason carries a lesson about pain that extends well beyond the knee, to the far-flung reaches of the body.

Let's look at what the procedure actually does. After you're under anaesthetic, an orthopaedic surgeon makes two cuts in the front of your knee to insert an arthroscope, which lets them see the torn cartilage and remove the jagged edge. But most surgeons want to shave the damaged area back to a smooth arc, which often means removing 15–20 per cent of the meniscus. How does that fix the pain? Here is where the uncertainty starts to creep in. The meniscus itself has a very limited nerve supply, so the tear often won't hurt of its own accord. Surgeons think the meniscus causes pain when the torn flap of cartilage rubs against, and irritates, the glossy, translucent tissue that lines the knee joint, called the synovium. Unlike the meniscus, the synovium has a copious nerve supply and is, therefore, well equipped to make your life a misery.

However, no one has shown conclusively that a torn flap rubbing on the synovium causes the pain. There is, in truth, a yawning abyss of doubt on this very point, one that has opened up around a statistic that should be a numerical touchstone for anyone considering surgery for pain. If you take a random bunch of people off the street in my age group — over 50 — one-third of them will have a torn meniscus. If that sounds like a lot, older people are especially prone to degenerative tears, which happen when the meniscus deteriorates, rather than when it's subject to an excessive force. But there is another, even more startling statistic. Sixty per cent of those people with a torn meniscus, a clear majority, have no knee pain at all. Take a moment to consider what that means, because it is monumental: having a torn meniscus is entirely consistent with being pain-free. In fact, there are millions of people walking around who don't even know they have a tear. Now, if you are someone with a torn meniscus, and you are in pain, and you are being offered a partial meniscectomy to treat the pain, learning that statistic may get you to pondering something: 'How can I become one of those over-50s with a meniscal tear who is pain-free, without having surgery?' This was my very own dilemma. Could I beat the pain and get my mobility back without an operation? But there was one question doing acrobatics over all the others. I knew from years of medical training that, after all these months, whatever healing was going to happen was done. *So why did it still hurt?*

The inner ping-pong of overthinking was getting me nowhere, so I'd booked to see one of the best in the business. Kal Fried is the go-to guy in my neighbourhood for people with pain that's standing between them and their bike, basketball, badminton racquet, or whatever movement gives them pleasure, health, or plain old freedom. We might see him for short-term, 'acute' pain, but Fried's special expertise is dealing with pain that lasts longer than three months, as mine now had, officially termed 'chronic' or 'persistent' pain.

Fried is a sports and exercise physician, and his CV is studded with the perks of the profession, namely, getting to hobnob with and lay healing hands on the athletic elites of the nation. He's been the club doctor for the Melbourne and Collingwood AFL football teams and worked with Australia's national netball team, the Diamonds, to name a few. But Fried is no sporting snob. His passion can be summed up in two words that apply equally to the sports star staring down a career-ending injury and the suburban pensioner with a bad back: pain literacy.

Fried is a champion for the cause that understanding pain is key to overcoming it. If that sounds orthodox, his approach is, well, quirky. His website is peppered with blog posts whose titles can lean towards the titillating. There is 'Sorry, but there are so many more than 50 shades of grey in pain' and the enigmatic 'A Sports & Exercise Medicine and Pain Revolution Copulation'. It once featured the following disclaimer:

> The many hours spent setting up and maintaining this website personally are not rewarded financially apart from the therapeutic benefit for me ... This may mean that I can avoid paying psychology counselling fees.

You may have gathered that the good doctor has a sense of humour. Which is perhaps one reason I felt little trepidation when the man himself appeared, to call me from my waiting room reverie.

Fried is 60ish, with a pate shaved smooth, olive skin, and piercing blue eyes bordered by crow's feet that dance, leprechaun-like, when he smiles. He was kitted in standard office gear of slacks and an open-necked business shirt, but, when we were both seated at his desk, I noted a road bike leaning against the wall behind him that had the telltale grime of many hard miles, so I knew there was lycra stashed away somewhere. When he spoke, his tone was equal parts confident and unhurried, but the message was charged with the urgency of his pain-busting mission.

I hadn't seen Fried for a decade, when he'd tended to a biking injury on my other knee, so we spent a few minutes on catch-up chitchat. But suddenly, with an impish smile and the flair of a magician, he thrust three playing-card boxes at me, each stacked neatly on top of the other. 'Grab onto those, Paul,' he said. The top box, he told me, was full of cards, but the other two were empty. I curled my fingers round the triple stack, which just fit in my hand, and held them for a few seconds before Fried asked for them back. Then he handed me the single box containing the full deck of cards. I held it aloft, all by itself, for a moment. 'Notice anything?' he asked, eyebrows raised and smile on high beam. I had. The full box of cards on its own actually felt heavier than the three boxes all together. Which, of course, it couldn't be. It had to be lighter, by exactly the weight of the two empty card boxes that were no longer in my hand. I did it again — same result, same mystifyingly weird feeling. Fried fixed his gaze on me with a triumphant air as he let the illusion sink in. I gathered that I wasn't the first patient to be taken in by this sleight of neuroscience when he told me how it worked with a fluency honed, evidently, by repetition.

The feeling of holding the single full pack between just your fingers, he explained, means the brain senses it as heavier than it really is, compared to grasping all three packs, which are felt across a broader area of the hand. 'It doesn't matter that you know that two are empty and one is full, and it doesn't matter how many times you do it, you get the same sensation,' he said.

'Which says to me that the brain is not only responsible for what we feel, but it doesn't get it right all the time.'

Remember that phrase, because it is critically important. I had knee pain, but it doesn't matter where it comes from; the problem with pain is that, when it drags on, it becomes, in many cases, a deception. A card trick played by the body that harnesses the dizzying resources of the nervous system to dupe the mind into thinking that the body is still hurt and that hunkering down to protect the injured part is the only option. Pain, like the Roman god Janus, is two-faced.

No one likes pain, but its first face is straight-up and, frankly, essential for survival. Along with swelling, redness, heat, and loss of function, pain is a cardinal sign of inflammation, the body's response to illness and injury. The fact that a bit of you hurts stops you from waving it around or running on it, and this slowdown in activity helps it heal. But swivel pain around and take a look at its second mug and what you find is an inscrutable poker face that has bluffed a goodly number of the world's population. Because the longer pain goes on, the less reliable it is as a marker of damage, to your knee, back, shoulder, wherever. For many people, this is a disturbing insight. How could pain, the consummate danger signal, whose dictates get us to the dentist for that rotten tooth and to the doctor for those broken bones, get it so wrong?

Over the last few decades, a gathering wave of research has shown that persistent pain can cause the nervous system to become ultra-sensitive to sensations coming from the affected area. Things that would normally hurt, hurt even more, and things that wouldn't normally hurt, like simple pressure, or movement, become painful. In practice — and I speak from experience — that means something as innocuous as a breeze blowing on a long-suffering knee can be misinterpreted as pain. It sounds like pure madness, but this devastating error is a by-product of perhaps the most exquisite example of physiology in the human body. It is called neuroplasticity, and it operates throughout the entire pain-sensing apparatus, from the nerves that take pain messages from the injured part, to their junction with the spinal cord, and even the areas of the brain that process the pain and make you conscious of it.

Neuroplasticity is something of a buzzword, thanks to Norman Doidge's bestselling tome *The Brain That Changes Itself*. Put simply, it's the ability of the brain to rewire itself in the process of learning new things, like how to play the violin, or in the process of relearning lost skills, like how to talk again after a stroke. The concept has upended decades of dogma that tell us the brain's connections are malleable as putty in infancy but become as hard-set as amber when we reach adulthood. Neuroplasticity is, without doubt, a bona fide celebrity in the world of neuroscience and is largely deserving of that hero status. But like so much that glitters, neuroplasticity has a shadow side, and there the lustre of precious metal is harder to spot. It is called maladaptive neuroplasticity, and, when pain stays put, it can mean stimuli that are non-nociceptive — not legitimately pain-producing — get mistakenly registered as nociceptive — they bloody hurt. This is what I call the pain mistake, and I can say its existence calls for a wholesale re-evaluation of the concept of pain. That's because understanding the pain mistake is a key to its undoing. My mission in this book is to supply at least some of that knowledge and show how it can put the brakes on many cases of persistent pain.

Along the way, we're going to meet a policewoman, crippled with pain for years after a devastating injury, who found the key to her recovery after hearing just two words. We'll hear from a legendary Harvard neuroscientist about how he discovered something strange the body does to keep people in pain. We'll meet a martial-arts champion who was pushed to the brink by pain from a freak accident and who recovered, without drugs or surgery, with something that many people already have in their lounge room. We'll hear from a man facing a knee replacement for osteoarthritis who, after a major heart operation, found a way to fix his pain without surgery. And we learn why a famous pain scientist felt nothing when he was bitten by one of the world's deadliest snakes but was in agony when he got scratched by a twig.

At the same time, we're going to tackle some crucial questions head on. How does understanding the pain mistake make it go away? If persistent pain doesn't signal ongoing injury, why is it there? Can you learn to be in pain, and, if so, can you unlearn it? Does exercise, which might seem to be making things worse, actually help pain? Are depression and anxiety only a consequence of chronic pain, or can they cause it, too? And we'll look at the latest research on techniques, including talking and physical therapies, that can wrest control of pain back into our own hands.

But we'll also grapple with a dire anomaly: despite their effectiveness, these therapies remain confined to the margins, almost unknown to the vast majority of people in pain. And there are a lot of those people. In the US health system in 2016 alone, an eye-watering $380 billion was spent treating musculoskeletal disorders including back, neck, joint, and limb pain as well as conditions such as rheumatoid arthritis and osteoarthritis. No fewer than one in five people live with chronic pain. Yet therapies grounded in an understanding of neuroplasticity are crowded out by pills, injections, and surgery, blockbuster treatments that can harm people by the millions. The US opioid crisis is exhibit A when it comes to pharmaceutical harms; overdose and suicide linked to opioid painkillers have been blamed for the first drop in US life expectancy in a century. But physical treatments also take a toll. Take spinal fusion for back pain. Rates doubled in the US in the decade to 2009, and the procedure generated costs exceeding $10 billion in 2015. Yet one in six operations leads to complications, including nerve damage. On top of that, experts are lining up to say that, as a treatment for back pain, spinal fusion is as good as useless — inferior to exercise, cognitive behaviour therapy, and physiotherapy. These pharmaceutical and surgical treatments are co-conspirators in the pain mistake. They are directed at a body part, not at the nerves that sense it. They seek to alter the anatomy, not the perceiving brain. They fail to understand the importance of maladaptive neuroplasticity and rewiring those oversensitive pathways.

All of which led me, during my meeting with Kal Fried, to ask myself a potentially life-changing question: was I a victim of the pain mistake? I had a rudimentary knowledge of maladaptive neuroplasticity from that earlier cycling injury, and wondered if it might be a factor in my current knee pain. But I'd seen the meniscal tear on the MRI scan with my own eyes. There was a highly visible, well-articulated causal chain leading from my overdoing it on the running track and basketball court, to the symptoms of a sore knee, to the diagnosis of a meniscal tear. The pain had lasted five months because, quite simply, I hadn't got it fixed. A surgical trim seemed the logical response. But I just couldn't get that 60 per cent figure out of my head; if having a meniscal tear was compatible with a life without pain, why couldn't that life be mine?

After the card trick, we got down to the nitty-gritty of what was going to fix my pain and, pressingly, whether surgery was part of the solution. Fortunately, people had been doing the hard graft on that very question. Soon after our consult, Fried emailed several articles, which I gave a thorough going over. And a little while later, I got another email from Fried. He'd just finished creating a learning module for the Australasian College of Sport and Exercise Physicians, and he invited me to take a look at it. It was full of Fried's signature quirkiness, including a slide of him riding up the torturous Alpe d'Huez, one of the steepest climbs on the Tour de France circuit. A speech bubble emanated from his panting lips: 'Pain is not an accurate marker of tissue damage, pain is not an accurate marker of tissue damage, pain is not …' There was also a slide with some statistics that bore out his point. It showed a skeleton with various pain hotspots circled in blue, including the knee, lower back, neck, and shoulder. Adjacent numbers tallied the abnormalities picked up on scans of those areas — but in people with no pain at all. They made for remarkable reading.

In the neck, 87 per cent of people aged 20–70 have a bulging disc in the spinal cord, with no pain. Thirty-seven per cent of 20-year-olds and 97 per cent of 80-year-olds have degenerative discs in their lumbar spine. With no pain. Forty-three per cent of over-40s have degeneration of the cartilage lining the knee joint. With. No. Pain. These are all abnormalities that, in people who do have pain localised to one of these areas, trigger a range of treatments to correct the anatomy. Yet there are literally hordes of folk walking around with the very same anatomy, and no pain at all. You could be forgiven for concluding that, in a generous chunk of people, the anatomy is not the problem.

I was now in possession of the most important tool anyone can have when deciding how to treat their pain: knowledge. I began to mull it all over as I worked my way towards answering that pivotal question: what should I do? I hope the stories in the coming pages, drawn from my interviews with some of the world's leading pain scientists and with people who have recovered from pain, will go some way to giving you knowledge that is relevant and meaningful to your own decision. In sum, this is a book about why pain persists — why it still hurts — and what we can do to overcome it.

Chapter 1

Who Got the Dux?

how the past prompts your brain to protect you

About halfway along the north coast of Tasmania, there's a flat sandy beach strewn with driftwood and fringed by banksia scrub. The beach rests on a cobble scree, which you can see at low tide, and, if you follow the stones east, they are soon riven by the mouth of the Forth River, where the banskia gives way to the gnarled, stooped forms of eucalypts that stand sentry on the estuary before it wends its way back to the hinterland. This is Turners Beach, and, when you trace the river for its first few hundred metres inland, you come to a squat concrete bridge where four lanes of the Bass Highway, the artery linking the sparse hamlets that dot Tassie's upper edge, cross the Forth.

On a fine summer morning just past daybreak in February 2015, Lauren was heading east on the Bass, riding two abreast on her road bike with her friend Tina. Lauren was in peak condition, just three weeks out from competing in the Melbourne Ironman, and, as they say in triathlon parlance, was 'on the taper', setting out on a paltry 120-kilometre trip as she eased back from some monster 200-kilometre rides. The day looked auspicious, and the two were exchanging breezy chitchat as they neared the Forth Bridge. Neither, it turned out, would get to the other side on their bikes that day.

'I just remember having this massive impact right across the centre of my back, then landing on the ground,' Lauren says. 'I expect the period between being hit and ending up on the ground was maybe a second. I remember thinking "I was hit by something", "This is going to hurt", and "Will I still be able to do the race?"'

The force turned Lauren into a projectile. She took out Tina, whose bike was left on the bitumen in the path of traffic, leaving Lauren stunned on the verge of the road. As she lay there, a car ran over the bike, and pieces of it flew past her. 'I asked Tina if she was okay, and she said, "Yes, move the bike, move the bike." I tried to get up. I tried to push myself up on my right arm, which was underneath me, but I couldn't. I couldn't elevate myself.'

Up ahead on the far side of the Forth Bridge, a truck had stopped and two men got out and were walking back along the roadway towards her. One of the men helped Lauren stand up and step over the guardrail to the safety of the berm, where she lay down and he covered her with a sleeping bag that he'd brought with him. There, prostrate on the stubble of roadside grass, she heard a refrain that would usher in a new chapter of her life, one composed, in roughly equal parts, of nightmare and revelation.

'He kept saying, "I'm so sorry, I'm so sorry. I just didn't see you."'

—

Lauren stands a compact five foot four, with sandy hair swept back in a ponytail, speaks in a no-frills Aussie twang, and has big grey-green eyes that, every now and then, glaze dull as the burden of what she's been through surfaces in the retelling. The crash left the 44-year-old policewoman with a shattered right elbow that a surgeon had to puzzle together with an array of metal plates and wires. Her right wrist was broken, and her ulnar nerve, which controls movement in the hand and wrist, damaged, all needing the surgeon's attention. Then, while she was in hospital, a sudden and massive bruising spread across her right hip and thigh; the fatty layer below the skin, the doctors said, had sheared from the muscles, and the space had filled with blood. It all hurt a lot, but, oddly, that broadside from a truck would soon take a back seat in the grand scheme of pain's stagecraft. 'I had pain I would have expected from the accident. I feel like I recovered physically in the right time frame, but I don't feel I recovered psychologically very well,' says Lauren.

It didn't help that Lauren had just mended from another reminder of her own mortality. In 2013, an intense pain in the back of her head stopped her in her tracks when she was doing tuck jumps at the gym. Within hours, she couldn't speak. A scan showed a brain haemorrhage had nearly ended her life. Now she'd been hit by a truck, on a clear morning, with flashing red lights on the back of her helmet and bike. You don't want to believe things come in threes. 'People were saying, "You've cheated death twice, you're going to run out of lives." It was almost like death was chasing me. That's how it was playing out in my head.'

It wasn't long after the crash that the terrifying dreams began. 'I suffered from nightmares about dying and would have constant thoughts and fears about it. If I walked over a bridge, I would picture myself falling off it, plummeting to the ground and then graphically see myself broken and deceased. I would dream about deaths I had seen in my work as a police officer, but I would picture myself as the deceased, laying in the wrecked car or laying dead in the bed.'

Tasmania's Motor Accidents Insurance Board reviewed the crash and asked Lauren to have a psychiatric evaluation. She had, said the shrink, nearly the full hand of symptoms of post-traumatic stress disorder, perhaps not surprising given her work history. She joined the Tasmanian police force at 22 and works at Burnie, a port town of 20,000 whose lifeblood is forestry and farming. She's dealt with everything from shoplifting to drug trafficking and murders. Over the years, as one of only two women in the Burnie Criminal Investigation Branch, she often found herself holding the short straw on sexual assault cases, whose victims were mostly women and sometimes children.

But when it comes to the forces that have shaped her psyche, Lauren swivels the telescope further back. 'I developed an awareness, quite early, that I had to do something special to get my parents' attention, because they were so busy. So if I won an academic award or was the lead in a drama production, they would come along to the presentation, and I had their interest for a brief period of time. I think this is partly what drove me to always want to win, be the best, do more than everyone else.' As a graduating cadet at the police academy, Lauren was runner-up dux and, bursting with pride, rang her mum to share the news. More than two decades on, her mother's reply still resonates: 'That's good, who got the dux?' Well, a 30-year-old former solicitor, as it happened. If anything, the approval vacuum just drove Lauren harder. At work, she climbed the ranks to become a police prosecutor. At play, she didn't just do triathlons, but the extreme, long-form version known as Ironman.

When 2017 rolled around, Lauren had healed enough to reignite her athletic ambitions. She decided to tackle an ultramarathon, and it was while training for that 100-kilometre slog that she got the first niggle of a pain that would, eventually, give her a third look at her own mortality. It started with a tiny stabbing sensation above her right knee that amplified until, when she was running, it felt like the tip of a knife piercing her skin. Lauren stopped running and saw an orthopaedic surgeon, who did an exploratory operation but found nothing. The pain got worse. 'I'd be sitting still, laying in bed, thinking about the injury stopping me doing what I love. I'd actually visualise the knife sticking into my skin then burning all the way up the side of my thigh.'

Lauren turned to the fickle wisdom of Dr Google, spending hours reading up on symptoms that might tally with her own to deliver that elusive, longed-for diagnosis. Her online exertions seemed to have finally paid off when she came across something called meralgia paraesthetica. An entrapped nerve produces a palette of pain ranging from tingling, cold, burning, and aching to lightning-flash spasms. Most people get symptoms on one side, but, if you delve into the fine print, you'll find that a small percentage of sufferers get it on both sides. Tenacious Lauren read the fine print, which may well prove that adage about the perils of a little knowledge. 'The next thing you know, the pain's on the other side, an intense burning. I'd put ice on my legs. I wouldn't have the doona on at night. I'd rest. I just didn't know how to stop it. Eventually, I was in pain in both my legs all day, every day, without respite. This went on for months.'

A doctor put Lauren on the painkiller Lyrica (pregabalin), which took the edge off her pain, and, after six months or so, she restarted a regimen of walks that progressed to running and cycling. The lure of the ultramarathon ever on the horizon, Lauren hit the pavement with a gathering intensity that, in the end, demanded full ownership of every fibre of her being. She had to do 25,000 steps a day, and, if it got to nine in the evening and she'd only done 20,000, well, she'd go out and run five kilometres. Back when she could only walk, she'd restricted calories to keep her weight down. Now whipped along by an almost satanic impulsion, she kept on that exigent diet. Her iron levels dropped, and she became anaemic. She developed a calcium deficiency. 'In October 2018, I became so ill that my hair was falling out, my teeth were breaking, and I hadn't had a period for some years. Despite this, I kept running every day until Christmas.'

That quasi-religious ritual was, however, about to undergo a dramatic conversion. Running down a gentle incline, Lauren planted a footfall that delivered a searing pain in her left buttock.

Weeks passed, the pain got worse, and the burning in her thighs, under control for so long, reared its head again. She couldn't sit at her desk, and had to stop work. The doctors injected cortisone into her buttock, upped the Lyrica, and put her on the opioid painkiller Endone (oxycodone). A scan showed a stress fracture in the tail bone, a diagnosis that provided initial but short-lived comfort, because the wrath of the pain gods had been unleashed, and, rampant and unrelenting, the pain spread to her buttocks, thighs, and lower back. 'There was nothing I could do to escape it, other than drug myself to sleep. Opioids didn't touch the pain, but they at least dulled my senses,' Lauren remembers.

'My PTSD symptoms escalated at this point, and I began having panic attacks. My nightmares were now incredibly intense, and, although all I wanted to do was sleep to escape the pain, I feared I'd die in my sleep. I would wake up gasping for air. I went from running marathons to being scared to walk to the letterbox in a matter of months. Doctors told me my fractures would have healed by this time and they could no longer explain my pain.

'I began seriously contemplating suicide, as I couldn't bear the thought that my life would be like this forever.'

It is said the darkest night is just before dawn, a glib platitude for some but apt for Lauren because, not long after that abysmal bottoming out of her will to go on, hope appeared in the form of three signs. First, despite the intensity of her pain, Lauren noticed that it wasn't disturbing her sleep, which was unusual. The second sign came on a day when unbearable pain drove her back to the doctor. Outside the surgery, she was doubled over and didn't want to leave the car. Finally mustering the wherewithal, she made it inside to find a new doctor, one exceptionally well endowed, it turned out, with both bedside manner and patience. 'I went in and just cried and cried, pleading for help, and, when I'd physically exhausted my entire body from crying, I didn't have any pain. I thought, "What happened with that physical release?"'

At this time, Lauren's life was insular, which is perhaps why a long-term plan to go to the Red Hot Summer music festival with her sister-in-law had never materialised. The 2019 show was to be a cracker, headlined by Suzi Quatro, and, as the anointed date edged closer, Lauren was getting well-meaning pressure to go, not least from her partner. Ring fenced with pain, the trip seemed more likely to be an ordeal than a pleasure outing. Eventually, Lauren relented and made her way down to the gig venue in the verdant, manicured, 19th-century surrounds of Hobart's botanical gardens. It's a magical place. There are lily pads on still water beneath a lone red bridge in the Japanese garden, and an oak canopy filters the hard southern sun to a dappled green on sloping lawns. Somehow, as the music surged and faded on the crisp air coming off the nearby Derwent River, Lauren's predicament receded from her mind. This was the third sign. 'For two days, I was with my sister-in-law, not thinking about it. I had hardly any pain. I came home thinking, "Why did that happen?"'

On cue, the demonic treadmill groaned back to life, and she was soon in pain again, waking up and interrogating herself: 'How much pain am I in? What's hurting? Can I go to work?'

She was researching surgery and emailing clinics, asking if they could help. 'I was willing to do anything. If someone said, "Have a spinal infusion," I would have done it. I was that desperate, totally obsessed with stopping the pain. But those three things kept ticking in my mind.'

How, Lauren wondered, had the pain disappeared for no apparent reason? She changed tack on Google and began scouring podcasts of people telling their pain stories. After a horse-riding accident, one woman had spiralled from keen rock climber to barely leaving a bed in her lounge room. The woman's research odyssey unearthed a term describing pain that, instead of coming from the injured part, was delivered to the brain by jangly, overactive nerves, a state she had been able to reverse. The term was 'central sensitisation'.

'That was the term that first prompted this,' says Lauren. 'Her injury had recovered, but her pain had never stopped. I said, "Wow, I wonder if that's what's happening to me?"'

Walk up Hospital Street in Johannesburg, past Mr Chips takeaway and supermarket, past the locals in bright garb perched on bollards and bins, nattering and gesticulating, and chart your course along a red-brick wall that steps ever higher with the rising ground, and you'll come to a solitary driveway. Peer along it and you'll see a low utilitarian structure with peeling white paint and louvre windows. Behind that, rising with a kind of austere majesty, is an imposing edifice whose windows evoke guard towers and where a cement spiral staircase, reminiscent of Frank Lloyd Wright's New York Guggenheim, clings to its side like a modernist barber's pole. This is the Old Johannesburg General Hospital, erected in 1939 by South African architect Gordon Leith, but with its architectonic glory days firmly behind it.

In the early 1970s, apartheid was in full force and, just a few short kilometres west of the hospital precinct, bitter resistance to government oppression was growing, soon to foment the Soweto uprising. There was upheaval within the hospital's surgical wards, too. In those days, if you had a stomach ulcer, it was common to have it cut out in the operating theatre — nowadays, you might take antibiotics or drugs that stop acid secretion. There were few nuances, either, if you had breast cancer. A total mastectomy — complete removal of the breast — was likely to be your lot, against removal of the lump with chemotherapy now. Those radical surgeries caused much pain for their recipients, a fact that didn't go unnoticed by a young medical student named Clifford Woolf as he flitted from bed to bed, tending to his charges.

Woolf was something of a gadfly, unafraid to harness his piercing intelligence to probe, Socratically, the often-opaque practices of his medical superiors. One day, he encountered the junior surgical doctor applying electrodes to the abdomen of a post-op patient, a newly minted technique called transcutaneous electrical nerve stimulation, TENS for short. Woolf quizzed him. 'Why are you doing that? How does it work?' The answer to the first question was, quite simply, to control the pain. The answer to the second was: 'Don't know, don't care, doesn't matter.' Pain, evidently, was part and parcel of the surgical experience. It was a message Woolf heard loud and clear from the outset. 'My introduction to pain was going into the postoperative ward and seeing all the patients in terrible pain and asking the surgical resident, "What's going on?"' says Woolf. 'He said, "What do you expect? They've had surgery. They have pain. They shouldn't complain." That's when I realised there was an enormous unmet need there.'

As that tumultuous decade drew to a close, Woolf's resolve to study pain firmed. He packed his bags for London, eventually finding his way into the laboratory of a certain Patrick Wall. Wall was a bespectacled, plain-spoken man with a Sigmund Freud beard and an anti-establishment drift. As a medical student at Oxford in the 1940s, he chaired the socialist club, before leaning further left to the communists, and finally embracing the teachings of Proudhon, famous progenitor of the anarchist war cry 'property is theft'. Wall had also turned his hand to writing fiction. In the mid-60s, he published a romp of a novel called *TRIO: The Revolting Intellectuals Organization*. It was about a psychiatrist who purged the frustrations of his highbrow clientele by unleashing them on a succession of ill-doers, including a corrupt businessman selling condemned antibiotics in North Africa. You get the impression Wall wasn't afraid to push back on the bedrock of conservatism, a trait that may have been instrumental in his rise to become perhaps the towering figure in 20th-century pain science. Wall, along with Canadian psychologist Ronald Melzack, had in 1965 published a radical paper that proposed an entirely new way to view pain.

At the time, there was an entrenched view among doctors,

and a good deal of researchers, that pain was much as Descartes had imagined it in the 17th century. In his book *Treatise on Man*, published in 1662, Descartes includes the bizarre image of a person with the pumped physique of a baroque Adonis and the grinning head of a mischievous cherub. The grin is probably a grimace, however, because the lad has his foot rather too close to the flame of a small twig fire. In his description of the pain pathway, Descartes refers to the fire touching the skin and pulling a small fibre, 'just as when you pull on one end of a cord you cause a bell hanging at the other end to ring at the same time'. The idea had persisted down the centuries: an injury like a burn, bruise, or broken bone, or an illness like arthritis or cancer, would simply pull the pain cord to alert the brain to the problem. 'It was thought of as a rather simple system, like you had a burglar alarm in you, and a bell went off when you were injured, and that was all there was to it,' said Wall in a 1999 interview with Australia's ABC radio. 'Doctors really didn't pay very much attention [to pain], they didn't listen to the patients, they thought it was a symptom and they had to be after the disease. They had come up with a very simple idea that you had pain fibres in your tissues, in your hands, in your gut, and so on, and that, when those pain fibres went off, you felt pain and that was it. So it was all the idea of a very simple, one-way system. I didn't believe a word of what I was taught in medical school, and nor did the patients.'

One reason for Wall's scepticism was the experience of doctors treating the horrific injuries of soldiers in battle, newly documented in scrupulous detail by the American anaesthetist

Henry Beecher after serving at a field medical unit at the Battle of Anzio. In early 1944, Allied troops in amphibious vehicles launched themselves at a thin strip of beach abutting reclaimed marshland near Anzio, a fishing port south of Rome. The subsequent defence of that beachhead, against merciless bombardment, left many soldiers dead and others with catastrophic wounds. One 'husky 19-year-old soldier', Beecher recalled later, was hit by a mortar shell that left a 'meat cleaver–like wound cutting through the fifth to 12th ribs near the vertebral column', lacerating his lungs, diaphragm, and kidney. The young man seemed agitated and in pain, but promptly settled when given a barbiturate sedative in a dose so small Beecher claimed it 'would not have controlled his pain'. The incident roused Beecher's suspicions of a curious disconnect between the gravity of the soldiers' injuries and their experience of pain.

He set about measuring pain in a study of troops, whose catalogue of trauma included long bone fractures, extensive soft-tissue injuries, and penetrating wounds to the chest, abdomen, and head. Only those thinking clearly enough to accurately report pain were enlisted in the study. For anybody who has been badly hurt or, like me, has treated victims of blunt and penetrating trauma, Beecher's results are almost unfathomable. Sixty-nine men, just under a third, had no pain at all. Fifty-five soldiers, a quarter, had only slight pain. 'Three-quarters of badly wounded men, although they have received no morphine for a matter of hours, have so little pain that they do not want pain relief medication,' reported Beecher in an article published shortly after the war's end. What was it about the theatre of war that led to so much apocalyptic injury causing so little discomfort?

A decade later, Beecher, working as an anaesthetist at the Massachusetts General Hospital in the US, decided to compare the experience of those soldiers with their civilian counterparts. And if you are wondering where in peacetime you might dredge up such a manifest of mutilation, look no further than your local hospital. On a daily basis, abdomens are incised to remove diseased colons and gall bladders. Chests are opened to patch distended aortas, and vertebrae are laid bare to be fused. Beecher selected 150 civilians having these and similar types of surgery, peppering them with questions about pain and recording their need for narcotics. He then compared notes with a matched group of 150 soldiers from his earlier, Anzio cohort. His findings seem to defy logic. Just under a third of the soldiers — 32 per cent — had requested pain relief. Of those paired civilians, an overwhelming majority, 83 per cent, had asked for painkillers. What could possibly make careful, controlled civilian surgery more excruciating than a war wound? Circumstances, concluded Beecher, are everything. 'In a situation in which a wound has great advantage and means escape from overpowering anxiety and fear of death on the battlefield (war wounds terminating military service), extensive wounds are associated with comparatively little pain. In a situation in which the wound connotes disaster (major surgery in civil life), lesser wounds are associated with far more pain than in the former situation,' Beecher wrote. 'The intensity of suffering is largely determined by what the pain means to the patient.'

Beecher's findings pose obvious problems for the Cartesian burglar alarm. If hopes, dreams, and expectations can interfere with the signal, the idea of an infallible one-way detector of tissue damage seems remote. Wall was impressed by Beecher's study. He also knew of other examples where psychology leaves its imprint on the malleable phenomenon of pain, ones much closer to home than the Anzio beachhead. 'What do you do when your child trips up and smacks into the pavement and starts crying — what do you do? You start distracting the child, you pick it up, dance around, coo. So distraction has been what has always been used,' Wall told the ABC. Amazing, too, how a well-chosen video can placate the unquiet child.

Folk medicine has other analgesics up its sleeve — literally, in some cases. Bang your elbow and the instinct is to rub it. 'Why do you rub a painful area, what's that about? Well, it turns out that that does inhibit, stop some of the pain messages,' said Wall. 'As soon as one began to look at the predicted mechanisms, they just didn't exist as a simple matter … there were lots of factors involved. Then we could start going back to the patients and ask, "Are these patients just responding to a single message coming in?" The answer was "Never", [there] were always other things involved.'

Those other things, concluded Wall and Melzack, were gates, admission ports in the spinal cord that could be slammed shut to deny entry to pain signals. How might these spinal gates work? Say you do actually bang your elbow. The impact will be picked up by sensors in the skin called nociceptors — a mash of the Latin *nocere*, meaning harm, and *receptor*, detective cells that relay signals. Nociceptors, which look a bit like the frayed denim on a pair of bell-bottoms, are actually the splayed ends of sensory nerves, on the alert for things that cause pain, like a whack, a scald, or an infected cut. Ping the nociceptor, and a narrow-gauge nerve called a C fibre carries the message up your arm to a relay in the spinal cord, where it jumps across to vertical fibres heading north. Ultimately, the ouch is registered in a part of the brain called the sensory cortex. On cue, you begin rubbing your elbow, vigorously, while jumping around and swearing like a navvy.

Now, all that rubbing activates a different type of sensor, one that detects pressure on the skin, called a mechanoreceptor. The rubbing message is promptly telegraphed along much fatter nerves called A fibres, which also converge on the junction in the spinal cord, before heading brainwards. Critically, Wall and Melzack proposed that both C and A fibres also touch on something else in the spinal cord: a small bridging nerve, or interneuron, that, much like a surly official manning a border gate, could stamp your passport and wave you through, or close the gate on your travel dreams. The traveller is pain, of course, and the messages carried by C and A fibres have very different implications for the trip. C fibres come with valid visas that open the gate to unimpeded travel. A fibres, on the other hand, make the border guard treat pain like a hippie with no outbound ticket. The spinal gate is slammed shut and passage denied. Rubbing your elbow, according to the spinal-gate theory, swings the door closed on pain before it can reach the brain.

The brain also gets a say in the state of the gate. For soldiers whose guts might be on display but who are overwhelmed with the delirious thought of leaving the war, or the child who is distracted into quietude, Wall and Melzack proposed a top-down mechanism whereby fibres descending from the brain could also close the pain gate. The theory had, and still has, much to recommend it, but the actual nerve junction in the spinal cord hadn't been teased out. It was believed to lie in a translucent, jelly-like bit called the substantia gelatinosa. This mysterious material lies in a rear-facing projection of the cord's grey matter that looks like an elk antler, named, aptly, the dorsal horn.

When Clifford Woolf landed in Wall's lab, his assignment was to seek out and bring to light those bridging nerves in the dorsal horn. His goal was nothing less than to discover how the spinal gate actually worked.

The main building at University College London is a neoclassical, Regency-era monolith fronted by a ten-columned portico and topped with a pediment that might have come from Rome's Pantheon. It houses the UCL art museum. Take a stroll through its sunflower-coloured rooms and you'll find an abundance of references to the links between art and anatomy, from paintings of standing nudes to classical Greek busts and an assortment of life drawings that have come, over nearly two centuries, from students at the nearby Slade School of Fine Art. Head south from the museum and, after a hundred metres or so, you'll come to the Anatomy building, whose facade continues the classical revival. Here, the pediment features the serpent-entwined rod of Asclepius, universal symbol of medicine and healing. The study of anatomy, it reminds us, is not just for art's sake, but also serves the human condition.

Inside the building at the dawn of the 1980s, as tartan-clad punks with cockscomb mohicans strutted the surrounding streets, the determined figure of Clifford Woolf was assembling equipment on a lab bench, busying himself with preparations for the spinal-gate experiments. His study subject would be the rat. Woolf began by rendering the animal insensible, carefully removing the upper portion of its brain under general anaesthetic, after which the anaesthetic could, humanely, be stopped. The rat could survive in this state for several weeks. Then he exposed the rat's spinal cord and located the dorsal horn, before staining it with a dye derived from the root of the horseradish. Now able to visualise the minute neurons within, he inserted tiny electrodes to record their activity. The difficulty of the task, however, would soon become clear. What Woolf found was a panoply of neurons that, not unlike the punks outside, seemed intent on asserting their individuality. 'Each cell was unique,' recalls Woolf. 'I published a paper with, I've forgotten how many, 20 or 30 cells. Each one was different. You could make a story about each one, about how they could potentially amplify or depress pain. But the reality was, we had no means at that time of dissecting out the circuits of the spinal cord. And I found that very frustrating.'

It wasn't long before Woolf decided to cut his losses on a task that, with the tools of the day, was mission impossible, akin to isolating each wire in a massed telephone cable to find out who was talking to whom. But he was by no means done. Was there another way, he wondered, to find out what all those fiddly neurons were doing, without meddling with them directly? His idea was, quite simply, to work backwards.

Think about what you would do if a young child got too close to a fire. Like any parent or other citizen mindful of society's smallest members, you are likely to get very jumpy indeed. Rightly so, but, even if disaster strikes, all is not lost, because the body has a remarkably effective built-in safety system. Should the toddler place their hand too close to the fire, it will be sharply withdrawn as their skin temperature rises, mostly before any nasty scorching occurs. And the system is almost foolproof, because it doesn't even need the child to know that pain signifies a forthcoming and very unpleasant burn. It's all a reflex that goes from the hand's pain-sensing nerve, to the dorsal horn of the spinal cord, and back out the motor neuron that controls the muscles of the arm. No thinking required; the brain is involved only as an afterthought. Pulling away a limb is, therefore, a faithful sign that a pain-producing stimulus is present. Why not, thought Woolf, use it to study pain?

The idea took root, and Woolf's focus on the anatomy of the lab rat shifted lower, down to the critter's hind legs. Right down, in fact, to its toes. Pinch a toe, or put it in hot water, and the toe will be snapped back quick smart. It's called a flexion withdrawal reflex, and, just like the toddler playing with fire, the outbound part of the circuit is completed by a motor neuron. 'I decided to move away from the spinal cord to look at motor neurons, and the reason for that was you know exactly what a motor neuron does. When it fires, it causes a muscle to contract,' says Woolf.

In humans, there are thousands of these motor neurons exiting the spinal cord. Initially, they travel bundled together in big, cable-like nerves, before going their own way and plunging into a part of their destination muscle. Woolf's plan was to dissect out individual motor neurons supplying the rat's leg, then pinch and scald his way around the skin of the leg to map out the areas where a painful stimulus would trigger the neuron and contract a muscle. The map for each neuron is called a receptor field. So Woolf began his methodical hind-limb cartography, and things started grandly, each motor neuron firing in response to a well-circumscribed area of skin being hurt. Then things got weird, or weirder — we are talking about rats with their brains removed suspended on a lab bench.

Some of those motor neurons would only fire when the burn or pinch was applied to a tiny area, often just a single toe on the rat's hind paw. Other neurons were less fussy. Much less. Some would fire when almost any part of the leg was stimulated, often with just the gentle stroke of a small paintbrush — no painful stimulus needed at all. It was a finding that simply didn't fit with how pain should work. 'I just couldn't make sense of it, because the withdrawal reflex is activated by a noxious stimulus. It's not activated by light touch, and also it has great anatomical localisation — you withdraw from the site where you're stimulated, you don't withdraw the whole body. So something was really strange,' says Woolf.

Indeed, if humans were equipped with these crazy neurons

Woolf had stumbled upon, you would expect to see kindergarten kids afraid of being hurt by a paintbrush, or pulling their foot away if some naughty preschooler pinched them way up on the thigh. None of which would serve pain's great humanitarian function of protecting the body from physical danger. As the months passed, Woolf continued his meticulous recordings. Outside, Britain was concluding the Falklands War, and Prime Minister Thatcher was trading barbs with Arthur Scargill, the trade unionist who led UK coalminers to their ill-fated mid-80s strike. Then, one day, something twigged for Woolf, and he had what he calls his 'aha' moment. 'I suddenly realised that all the motor neurons that had large receptor fields and a low threshold of activation by innocuous stimuli were recorded at the end of the day. Those that had the tiny receptor fields that required intense noxious stimuli were recorded at the beginning of the day.'

What possible difference could the time of day make to how much pain you have to inflict, and over what area, to get a motor neuron to fire? This is where the deductive powers of great scientists come into their own. Those unfussy motor neurons, with a hair-trigger response to a brush anywhere on the leg, were not a newly discovered category of cell but, rather, had been made that way. Made, not born, by Woolf himself.

Woolf is in his 60s, wears dark horn-rimmed glasses and an open-necked shirt, presents a clean-shaven face and scalp to the world, and speaks with a light South African accent in a soft, measured register. Even after nearly 40 years, however, the excitement of this moment is clear, and, as he describes what happened, his tone shifts from patient and professorial to animated and, well, proud. 'That "aha" moment was when I realised I was changing the spinal cord during the course of my experiment,' says Woolf.

What could Woolf possibly have done to change the nature of the nervous system itself? He had, very simply, inflicted pain. Each day Woolf might isolate and record from 15 or 20 motor neurons, and, with each one, he would go through the same rigmarole of inflicting a burn or a well-calibrated pinch to an atlas of regions over the rat's toes, hind foot, and leg. Over the course of the day, the animal's paw had become injured and inflamed, and the stimulus needed to set off the withdrawal reflex had become smaller and more widespread. The mechanism, Woolf believed, was a bunch of newly sensitised nerves in the spinal cord, and those rejigged pain-sensing nerves were, in turn, creating a bunch of trigger-happy motor neurons.

But Woolf needed more proof. So he decided to track the activity of a single motor neuron over the course of a day. As the sun drew its arc across the London sky, Woolf went through his routine of pinching, burning, and stroking the rat's paw and leg, all the while focusing on the contraction of one motor neuron. As the day wore on, the neuron's receptor field expanded massively and its threshold for firing plummeted. It was just as he had predicted. Yet still Woolf was troubled. Could the whole phenomenon be caused by the injuries lashing the skin's pain receptors into a super-sensitive, rapid-fire state? Woolf injected local anaesthetic into the skin around the injured foot and repeated his poking and prodding. Even with the damaged area numbed, the motor neurons fired to ever wider and smaller painful stimuli. The nerves of the spinal cord had undergone a profound change. 'It was the moment I suddenly realised I was not looking at a fixed nervous system that had defined properties, because that was the mindset then. In fact, the nervous system was changing, and I was the instrument for the change by virtue of producing injury,' says Woolf. 'I was able then to show that this change was triggered by the peripheral injury, but, once it was triggered, it remained. It was a memory, if you like. And that was the discovery of central sensitisation.'

The implications of Woolf's finding were staggering. Repeated injury was altering the nervous system to become so sensitive that even harmless stimuli, the caress of a soft brush, could trigger a pain response, and do so across a broader area of skin as time passed. If the mechanism held for humans, it might explain pain that persisted and spread after a slow-healing injury. It could also upend the treatment strategies of a planet-load of doctors whose patients were in chronic pain; perhaps the source of pain was not where those people were feeling it — in the back, knee, neck, shoulder, wherever — but in the overexcited nerves of the spinal cord?

Woolf wasted little time in sharing the data with his esteemed mentor Pat Wall. The response from the great man was not what Woolf had hoped for. 'He thought it was interesting, but he said it was completely the opposite of what his whole work was, and he said that he felt that he'd allow me to publish it without him being co-author,' says Woolf. Wall simply wouldn't put his name to the work. So Woolf went it alone, submitting the paper to the world's preeminent scientific journal, *Nature*, with only his name on the manuscript because, he says, 'there was no one else who'd believe me.'

At least one reason for the general scepticism was that, despite many groups doing this kind of research, no one had seen anything similar. Why should Woolf have happened across a phenomenon that, with all those labs fiddling with rat limbs and motor neurons, had not been made apparent elsewhere? The answer, it turns out, is about ethics. To comply with ethical guidelines, all rats had to be anaesthetised before the researchers set to work with their scalpels and needles. But Woolf was focused on a reflex, and he knew that the anaesthetic gases needed to keep a rat asleep didn't just work on the brain. Carried by the bloodstream, they also enter the spinal cord and block the flexion withdrawal reflex. Which is why Woolf had made his animals decerebrate — removing their brains to both comply with ethics and preserve the withdrawal reflex. 'Everyone was studying the spinal cord in the presence of general anaesthetic, so they were suppressing the phenomenon I discovered. They couldn't see it.'

In the end, *Nature* published the paper, in December 1983. Woolf, who is now Professor of Neurology at Harvard Medical School, was rapt. 'I thought, "Wow, I've got a *Nature* paper." I thought I'd discovered something really important.' But if he was expecting a tsunami of acclaim, it was not to be. Pat Wall would get hundreds of requests for reprints from other scientists when he published a paper. 'Over the next year, maybe I got three,' says Woolf. 'It had zero impact immediately, nothing.' New ideas, of course, take time to remould the universe around them, and Wall eventually came round, reflecting ruefully on his decision to avoid putting his name to that paper.

'A year later, he said that was the biggest mistake he's ever made in his career.'

The focal point of Launceston in northern Tasmania is the Cataract Gorge, a tree-lined gash in the local dolerite rock plunging down to an emerald river that widens to become a swimming hole between the cliffs, packed with bathers on hot summer days. There are man-made aesthetics, too, with a wrought-iron bridge spanning the river mouth, and quaint Victorian and Edwardian houses clinging to the less precipitous slopes at the gorge entrance. In fact, Launceston proper is loaded with gorgeous period architecture, making its real estate a hot-ticket item for mainlanders. The Hotel Grand Chancellor is, sadly, in a separate category. The hotel, just a short stroll from the gorge, is an '80s pastiche of faux mansard roof, dormer windows, and Louis Quinze gestures in render on cladding. Fortunately, Lauren had not made the trip for the hotel's visual amenity but, on an overcast day in March 2019, to hear a talk by a group called Pain Revolution.

Pain Revolution was founded in 2017 by Professor Lorimer Moseley, a physiotherapist-turned-academic dubbed 'one of the most creative pain researchers alive' by psychiatrist Norman Doidge. Pain Revolution's mission is to empower health professionals and the general public with the latest pain science.

One way it does this is by dressing a gaggle of pain experts in pink-and-yellow lycra and putting them on bikes for an outreach tour — they ride the byways of rural Australia to hold a series of town hall meetings.

Lauren, at this time, was not doing great. The dark cloud of pain was messing badly with her life. She couldn't work and spent her days listening to podcasts of people's pain stories. Her headspace, usually decluttered with the brisk broom of exercise, was brimming with anxiety, and she was avoiding people. Her right arm was scarred, and she couldn't straighten it. 'When things got on top off me, I'd run, and I couldn't do that. I felt trapped. I'd look at myself every day and see that my body was different from other people. At the gym, everyone who was pushing weights had their arms in the air straight. Mine was bent, and I felt like I was broken.'

It wasn't all gloom, however. An exercise physiologist had Lauren walking each day, a kilometre at first and then, if she felt okay, a little further. She was stretching, doing light weights, and 'belly breathing', also called diaphragmatic breathing, which can ease pain, although it's not clear how. In her online wanderings, she'd come across Moseley, a charismatic and compelling speaker, giving a talk, and decided to make the trek to Launceston. So it was, body and mind on empty after four years on the punishing roundabout of pain, that Lauren walked past the neat box hedges in the Grand Chancellor forecourt and under the ersatz Doric columns of its portico, into the foyer. And stopped dead. Lauren's anxiety had taken hold, and, uncertain how many people might be inside, she couldn't face the seminar. But she didn't do an about-face. Lauren had spotted a woman behind a help desk who was, it turned out, from Pain Revolution. Her set-up was far less intimidating than the cavernous, glitzy space where the talk was happening. The two ended up chatting, much of it touching on central sensitisation and, as Lauren listened and other folk with pain milled around and shared their stories, something shifted.

'This woman was saying to me things like, "You can feel pain long after the injury has healed," which was something I didn't understand prior to that. I didn't understand that I could keep getting those pain messages long after the fractures had healed.' The woman had laid out some brightly coloured fact sheets, and Lauren leafed through them. They contained potent messages. Persistent pain, one explained, is a poor indicator of damage to muscles, soft tissues, and nerves. Another called pain a 'danger detector' that can switch on when danger is merely perceived but not actual. Yet another flyer told of how the hyped-up pain system can, with the right tools, be toned down. The messages resonated deeply with Lauren, and, as they sunk in, a weight lifted. 'It was a big relief. I thought, "Now I can stop looking for a diagnosis. Now I can take some control back." Prior to that, the control was in someone else's hands. You're always waiting for someone to give you a magic pill. This was something I could do. I could help myself. It just all clicked. I thought, "Yep, this is the direction I've got to go in. I'm sick of medication."'

Working with a psychologist and an occupational therapist, Lauren assembled an armada of strategies and got to work. 'I learned that my body is working hard to send me signals because it's protecting me, and that I need to retrain it to not send me so many signals. My role was to try and desensitise the system over time, gain understanding that I wasn't broken.'

Lauren kept walking, striding out further each day. She changed the language of pain at home. Instead of asking, 'How's your pain today?' her partner would say, 'Are you comfortable?' On bad days, when the pain flared up, she would use self-talk: 'My body is being protective, and it's okay.' She addressed her body directly: 'Thank you for protecting me, but I don't need you now.' She used breathing strategies, visualising the pain — sometimes as the letters of the word *pain* itself — leaving her body as she exhaled. She adopted an attitude of acceptance, remembering something her occupational therapist had told her: 'Don't fight it. Acknowledge it and work with it.'

Yet it was never plain sailing. 'Once you get an understanding, it doesn't just fix it. Once everything clicked, I thought, "Now I can fix it." I stopped taking Lyrica and went for a run, and everything intensified. I thought, "Oh no, I'm wrong." It took me a while to accept that it takes a while. You can get disheartened. The message from the Pain Revolution is that recovery is not a straight line.'

Well-meaning family, friends, and colleagues were always making 'helpful suggestions'. Perhaps try this surgeon, that doctor; maybe take cannabis oil. Lauren decided to write them a letter, sending it to her daughter, her boss, and her friends. 'This is what's happening to me. I need psychological and physical work to recover. Please come along for the ride,' she wrote. On the way, Lauren also got plenty of messages herself from health professionals, some distinctly unhelpful. After the stress fracture, one told her that, 'fractures are just all pain I'm afraid'. This, says Lauren, 'set my mindset that I was in for lots of pain'. Another said, 'you've been through so much, this little injury shouldn't bother you' — 'this was dismissive and made me feel like I shouldn't complain about my pain.'

But one message struck a perfect chord. It was, very simply, 'your pain is real'. 'This was a big onc! It was so important to hear this statement,' says Lauren.

When I first got in touch, Lauren sent me some notes to explain what had happened to her, some of which I've drawn on here. It was a seven-page Word document that, as I knew I'd be interviewing her shortly and getting the story from the horse's mouth, I thought I'd just skim quickly. So I was hurried and a bit impatient when I started reading what I thought would be a formulaic account of her injuries, scans, and visits to the doctor. By page three, I was blubbering and heaving in big, cathartic draughts. The crash, the devastating injuries, the PTSD — all primed me, but there was something about a few recollections of childhood that pulled the trigger.

Lauren never doubted her parents loved her, but, too often, they weren't there for her. In primary school, her mum and dad were mostly up and gone before she was out of bed, and would leave lunch money on the bench. She can't remember ever taking a homemade lunch to school, which made her the envy of other kids because she had money for the canteen. But Lauren saw it differently. 'I was so jealous of kids who had cut lunches that

I once got into someone's bag and ate their beautiful-looking sandwiches and got caught and in trouble for it.' When she was in grade five or six, Lauren's mum was supposed to pick her up after sporting practice, but never showed. Someone who lived by the school noticed Lauren sitting there at 5.00 pm and got on the phone. 'When my mum came, it was like it hadn't happened. There was no apology, I was just dropped at home and she went back to work.' And then there was the 'who got the dux?' incident after graduation. It was, Lauren says, 'as if my achievement wasn't the best, so it wasn't important, although I know she never intended it that way'.

I'm sure I am not alone in recalling memories of an absent parent and feeling an empathetic pang for Lauren. It is tempting to draw neat psychological conclusions. Lauren's hunger for her parents' attention spurred ever-greater efforts to be noticed, approved of, and loved, culminating in a near-obsessive need to succeed. Sporting triumph got entangled with Lauren's self-worth, and failure was the bogeyman that could crumble her self-esteem. Pain posed a threat to winning and to the daily exercise that was a balm to Lauren's psyche, and so it morphed from a normal and manageable part of recovery into a danger signal that fed upon itself.

Plausible? Maybe, but the theory is, of course, overly simplistic, improbably linear, and, almost certainly, unprovable. What is increasingly clear, though, is that attitudes are central to the persistence of pain, no matter what their origin. 'Catastrophic thinking', for example, where pain is taken to be a major threat with little chance of resolution, is a potent predictor of chronic pain. Which leads to a part of Lauren's extraordinary story that left me wondering.

Lauren is recovered now. She's working full time and even did the 700-kilometre Pain Revolution ride through the countryside of south-eastern Australia in March 2020, speaking on panels about her experience to an audience of people whose shoes she has truly walked in. Her story is one of hopes and fears, dreams and disappointments, glimmers of optimism and shattering lows, all influenced by the words of doctors, messages on a flyer, and even the strains of Suzi Quatro's 'Can the Can' drifting over an ornamental garden. These are, when it boils down to it, a bunch of thoughts and feelings. Yet they carry an immense power — nothing less than the ability to dial pain up or down. But central sensitisation and the complex factors that amp up the nervous system involve wired-in changes to nerves. Which raises a perplexing question: can thinking really change your nervous system? And if the answer to that is 'yes', and the result is pain, can thinking change it back again?

Chapter 2

The Rabbit Hole of Who You Are

correcting the brain's image of the body

Weekends can be stupendous, but, not infrequently, they involve some of the dullest minutiae of our often-pedestrian lives. On a dreary Saturday morning in March 2011, as brown autumn leaves drifted from the plane trees that line the streets of Melbourne's Elwood, a schoolteacher named Carl was on the floor of his bedroom, attending to a problem in the latter category. His bed was oldish, a bit rickety, and had reached the point where, if Carl or his partner merely rolled over in their sleep, it would groan loudly and disconsolately, often waking them up. The job of repairing the thing had risen up Carl's mental to-do list and was now at the top.

Carl is a big guy. He's six foot six in bare feet with brooding dark eyes, close-cropped, receding hair, a wraparound goatee, and a facial expression that defaults somewhere between serious and deadpan, which would be intimidating if not for the intermittent one-liners he tends to follow up with crackles of laughter. Suffice to say, it was quite a bend for Carl to get down to eye height with the bed crossbeams, upon which he was intent on fitting some reinforcing brackets. There was a serious geometry mismatch between man and bed, and something had to give.

'I spent a lot of the day crouched over, leaning over,' remembers Carl, who's now just shy of 50. 'I was on my knees, and I sat up, and my back felt quite tight in a spot. I wanted to move, and it sort of felt like something was stuck, and I sort of forced myself to straighten my back. I was on my knees and went up, and I felt this almighty pop, and a shot of pain down my leg. I thought, "This is not good."' The pain landed its blow behind Carl's left kneecap. 'It had an electric feel to it, and it was hot and burning, like a nail being driven into the back of your leg.'

He got checked out at the emergency department and was sent home on painkillers to rest and take time off work. But when things didn't get better, Carl booked in to see an orthopaedic surgeon. 'He said, "Look, I don't really want to operate. With a bit more rest it will go away. I sort of got the impression he was saying I was being a bit of sook.' Nonetheless, Carl took the advice and gave it more time, but, by August, things were getting dire. He was on strong painkillers and having trouble urinating. He was numb on the outside of his left leg, and his left foot had become floppy. 'I was reluctant to go back to the surgeon because, I guess, of what he'd said the first time. But then I realised I was in trouble. I could barely get to the toilet.

It wasn't getting any better.'

Carl dragged himself back to the surgeon, who now stepped things up and arranged an MRI of his back. It wasn't good. A bulging disc at the lower end of Carl's lumbar spine was compressing nerves that control the bladder and handle movement and sensation to the leg and foot. What was needed, said the surgeon, was an operation. Those protruding bits of disc would have to be trimmed to relieve pressure on the nerves. Unfortunately, thanks to the back's anatomy not having evolved to make life easy for surgeons, accessing the disc is a little tricky. It often means cutting out the bony roof over the spinal cord formed by the laminae, which join to form what looks, on cross-section, like a gothic arch. Doing away with the arch also takes pressure off the nerves as they peel away from the spinal cord and head to the legs.

The surgery was scheduled, and, in September, Carl underwent a laminectomy and discectomy at the level of the fourth and fifth lumbar vertebrae. As is turned out, the precise state of Carl's offending anatomy was only hinted at in the scans, but confirmed under the learned gaze of the surgeon after stripping away his back muscles and chiselling out the vertebral arch. 'He said it was one of the worst he's ever operated on. There was one millimetre or so for the nerves to move through between the disc and the bone,' says Carl.

The surgery went to plan, and Carl was discharged home. But a week and a half later, things went pear-shaped again. Sciatic pain, shooting down Carl's leg — a cardinal symptom of nerve compression — had returned. But the portion of disc encroaching on the nerve and the bony capsule overlaying it were both gone. So what was causing his pain? Carl dutifully returned to the surgeon. 'He was reluctant to believe me and was telling me I was panicking for no reason, but I was pretty persistent,' says Carl.

He went in for another MRI scan, and, yet again, it was bad news. The way disc protrusions work is that the jelly-like central portion of the disc, the nucleus, herniates out through the fibrous outer ring, the annulus. In rare cases, even when an initial herniation is excised, the jelly can, undeterred, poke through again. Carl, who has a science degree and has worked as a university researcher, ran the maths on this happening. The chances were just 5 per cent, but that's exactly what had happened.

Two weeks later, Carl was face down on the operating table for a second time, having what's known in the trade as a revision discectomy. Afterwards, the surgeon was sanguine on Carl's prospects. 'I remember him saying to me, "You should be right now, because there's nothing left in the disc. It's pretty well empty."'

The body, of course, has its own rhyme and reason, and the prognostications of doctors are often up there with the bad old days of rain forecasting before the weather satellite. Prone, that is, to being a little askew. 'I'd had sciatic pain, but I'd never had any lower-back pain before any of this,' says Carl. 'After that, the sciatic pain disappeared, but I had, oh, just massive lower-back pain, and I could barely stand or sit. The operation was in October, and I ended up going back to work at the start of the following year in January. I was only doing half-days, but, even using fentanyl patches and tramadol at once, I could barely get through a four-hour day. I was just in immense pain.'

Now Carl had to run a different kind of maths, one that weighed the odds on his future. The calculation would be life-changing. 'I realised then that I wasn't going to be able to work. I was just going to have to resign.' He took a redundancy from the university and tried to strengthen his back with physiotherapy and Pilates. 'Things never really changed. I couldn't stand or sit for more than ten minutes without a lot of pain. I'm talking about probably eight-out-of-ten pain. I've had gallstones, and nine or ten is when you're writhing around on the ground trying to find a comfortable spot. An eight, you can still move, you still sort of have some control. But it was pretty bad.' It was at this time the toll on big stoic Carl really began to show. 'I know there would have been people worse off, but I remember lying in bed and thinking to myself, "There are definitely some things worse than death." I was thinking, "If this is what the future looks like, it's not good."'

Carl now climbed aboard a merry-go-round of treatments that would last two years. Their effectiveness was, to say the least, variable. He saw a pain specialist who injected a glucose solution into the space around the spinal cord, a treatment that has shown promise in some studies but whose mechanism is unknown. It was no panacea. Then he had something called radiofrequency ablation, which uses electrical current to heat up and knock out nerves that carry pain from the vertebrae and discs to the brain. This delivered some, but short-lasting, relief.

In desperation, Carl flew to a clinic in the United States to have stem cells taken from the marrow of his hip bone and injected into the remnants of the damaged disc. It's still being tested in clinical trials but got Carl's pain down to a more manageable five or six out of ten. He could not, however, work full-time.

'By that stage, I'd sort of figured I'd tried all the possible physical or medical interventions that were going to make a difference. And that's when the GP recommended the Barbara Walker pain clinic.' Carl didn't need too much arm-twisting, but, when he went for a look-see, the clinic offered him something that wasn't on any pain management checklist he'd ever read. 'They recommended this three-week pain course where you come off all your medications and learn how to deal with it. I was extremely sceptical about that. I just wasn't prepared to go without the painkillers.' Rolling around in the back of Carl's canny cranium was a very fair question: what kind of clinic would *stop* painkillers to make your pain go away?

In Melbourne's inner north, there's a big old building that looks to have erupted from the earth along a fault line of architectural motifs that span two millennia. Its Romanesque arches could have sprung from Hadrian's Villa and its crowning dome echoes the renaissance mastery of the Florence Duomo. Fittingly, its vaulted halls have witnessed defining moments in Australian history, from the Melbourne International Exhibition of 1880, for which it was built, to the opening of Australia's first parliament in 1901 and, not least, the matriculation exams of thousands of school kids, one of whom, a century after it was built, was yours truly. Wander east across the expansive Carlton Gardens out front, past the fountains with their Victorian-era statuary, the translucent green sprays of the English oaks, and the massive serpentine buttress roots of the Moreton Bay figs, and you'll find a cream-brick box of a building in a complex that makes up St Vincent's Hospital. If the Royal Exhibition Building is life's lofty grandeur, this brick box is its hardscrabble foil, because inside is a fragile coalition of souls for whom misery has been distilled to its essence.

This is the Barbara Walker Centre for Pain Management, named after the wife of late Melbourne businessman Ron Walker; her own battle with chronic pain prompted the pair to fund the clinic. The centre has, for the last 21 years, been led by pain-medicine specialist Jane Trinca. 'Most people have had pain for about ten years,' says Trinca. 'Often, they've been to see lots and lots of doctors and physios and chiros and all sorts of people and they've still got their pain. They're on lots of painkillers, they've reduced their work and family relationships, they've got depression. Why have they come to see us? Well, they often feel like this is the end of the road. Sometimes it's because their doctors have just said, "There is nothing more I can do for you, go to the pain clinic."'

Trinca's route into pain was roundabout. As a young anaesthetist in the mid-1980s, she was on the hunt for a staff job in a public hospital, a rare prize, and finally landed a three-month stint when somebody went on sabbatical. But there was, she was told, a catch. She would have to look after pain patients, legendary in medical circles as some of the toughest customers to treat, and a group many doctors went out of their way to avoid. Every Wednesday morning, Trinca would do a 'blocks' clinic, injecting local anaesthetic into a variety of body parts to treat pain syndromes. But as the weeks passed, her focus on the hard-nosed job of numbing nerves came adrift as she noticed some curious traits among her brood of patients. 'I'd see all these really strange patients on the ward who seemed to do really weird things,' says Trinca. 'I remember a woman wailing with chest-wall pain but no apparent cause. I gave her a very tiny, sub-therapeutic amount of intravenous analgesia, and her pain disappeared immediately. I almost wondered, "Should we be using placebo injections?" I thought, "There is this other element to the pain."'

A colleague suggested Trinca feed her growing interest in pain with a trip to the 1990 conference of the International Association for the Study of Pain, in Adelaide. She went and, in the process, met none other than Patrick Wall, Ronald Melzack, and 'new kid on the block' Clifford Woolf. Trinca was 'blown away', and the experience strengthened her suspicion that pain strategies aimed at the mind as well as the body could improve the return on investment. That ethos now grounds the Barbara Walker Centre's approach, but has led to policies that aren't always popular with people attending their flagship program, called START.

'They have to agree that they're going to give up all their medications that are used for pain,' says Trinca. 'So if they're on opiates or Lyrica or antidepressants for pain, they go.' The edict is too much for some, whose winnowing gets the thousand or so patients referred to the Centre each year down to the hundred deemed ready to occupy one of its few, sought-after places. But what method could there be in the madness of stopping painkillers for pain?

The big one is something called opioid-induced hyperalgesia. Opioid drugs like codeine and morphine put a lid on pain at the start, but they have a flip side: after a while, they can wind up the pain system, too, making it ultra-sensitive. A little like electric wiring with its insulation, pain nerves are sheathed in glial cells, which were always thought of as simple physical support structures. We now know they pack a hefty bag of tricks, one of which is to pump out inflammatory proteins — under the influence of opioids. The result is pain-sensing nerves that fire way more than necessary, dialling up pain.

But there is another, altogether-different reason why opioids get the chop under the Centre's no-nonsense regime. 'Some people on opiates think they're helpful because as soon as the level goes down in their bloodstream they feel "not right",' says Trinca. 'Their pain seems to come back, then they take more and they feel okay again. But often this is just the withdrawal of the medication. The drug itself has an effect so that, when you're not having it, you don't feel quite right, and, as soon as you have it, it's not like you feel high, you just feel normal again.'

The drug has, for some, become a crutch, worming its way into their psyche as essential to the daily plan when really it is an optional add-on. It's a bit like Disney's flying elephant Dumbo. He thought he needed a magical feather to fly, but finally discovered it had no special powers — his big ears were responsible — and he could fly just fine without the feather. Deemed necessary for flight, Dumbo's feather actually impeded it. Drugs are not, however, the only crutch Trinca's team must prise away to expose raw minds to the job of pushing back pain. 'There will be people who see their massage therapist every second day, or are so dependent on hot packs they have skin burns. They might have a special pillow, or they need three pillows under the head and one under the foot. There are people with a special chair, and they can't sleep any other way but in that chair. They have to give up all that stuff. They have to give up having lots of rests and lying down. We see some people who have become basically bed-bound, unnecessarily, because of their pain.'

On any given day, the Centre is a whirr of steady activity, pain-medicine specialists giving lectures, physiotherapists with their charges stretched out on mats, and, behind closed doors, people taking a pickaxe to their mountain of preconceptions about pain, guided by psychologists. Trinca is fastidious about ruling out causes of pain with a clear-cut treatment, testing for everything from a misplaced screw after a spinal fusion to the dreaded shadow of cancer. Once that screen comes back clear, she's ready to rattle the cage that keeps the solution to pain in the part of the body that hurts.

'You can't really diagnose a lot of pain conditions on an X-ray or an MRI. You can see pathology, yes, but that doesn't necessarily mean it is causing the pain,' Trinca tells me. 'I often say, if we went out into Victoria Parade and we took an MRI of everybody's spine, we'd find some people have pristine-looking spines but they complain of horrible pain. And we'd find that others have horrible-looking spines but they don't really have pain. So to go down the avenue of investigation after investigation trying to find something that's abnormal and then taking it out by a surgical procedure just doesn't work for many. We tell them about the results of back surgery. The general message is that, whilst there are things that obviously need surgery, absolutely no doubt, a lot of people are very disappointed. They have multiple surgeries, they sometimes have short term relief, but then they end up worse off.'

Trinca explains that spinal-fusion surgery to stabilise vertebrae, by knitting them together, sometimes with bone harvested from the hip, can transfer load to vertebrae above and below the fusion. That extra load can overwork muscles, which respond with spasm and pain. The figures bear this out. One review found a fifth of fusions needed reoperation within five years, and, in up to half of those, adjacent vertebrae had been damaged by mechanical stress. In 2018, Australia's Choosing Wisely campaign advised that, quite simply, spinal fusion should not be done in people without signs of nerve compression, or whose pain wasn't clearly caused by cancer or infection in the vertebrae, which can make the spine unstable.

All of this careful preparation leads to an essential realisation on the journey out of pain. It establishes that it's safe for the person to move. Safe to come out of the cone of protection put in place by a fear, often years-long, that movement can only cause injury. That person is now ready for the next crucial part of the program. They are ready to pace.

Pacing is the gradual reintroduction of movement to bodies that have shunned it. The team captures the first foray of their pupils on camera as they brave the long corridor at Barbara Walker, and it isn't pretty. 'They usually walk quite slowly, and they've usually got terrible posture,' says Trinca. After this desultory baptism, each person is presented with a little blue-and-white device they clip to their clothing. It is a mini alarm clock that will become their drill sergeant, calling time on whatever movement they have trouble doing — lifting, lying, sitting, standing — limiting it to as much as they can manage before the pain sets in. Then, with each passing day, the pocket tyrant has them do it a little longer.

Trinca has blonde hair cut in a short bob, wears '50s-style glasses that could have been borrowed from Catwoman, and has the glowing complexion of someone decades younger than her 65 years. That ruddy good health comes despite her own adversity: breast cancer, a messy divorce, and being a single mum all hit in her late 30s, just as she was entering the world of other people's pain. She's from a medical family — her dad was a GP anaesthetist — and she's a real doctor's doctor, committed to the daily grind and satisfied with the sporadic reward of her patients' triumphs. For my benefit, she's asked a colleague to rake together stats on every patient who's been through the START program since 1998. It's stuff that no one sees, that's not trumpeted on a slick corporate website, but it's hard not to be impressed.

Of 857 patients, 795 (93 per cent) finished the three-week course, and Trinca has 12-month data on 450 of them. The youngest was 17 and the oldest 90, with a mean age of 45. Two-thirds were female. They had dealt with pain for an average of nine years, ranging from less than a year, all the way to a hellish 67 years. Over half the patients had pain at two or more sites, the most common being the lower back (62 per cent), leg, knee, or foot (55 per cent), shoulder, arm, or hand (50 per cent), chest and upper back (47 per cent), and neck (46 per cent). The burning question is, how did they do? The answer is spelled out in metrics that might seem mundane but, for each person, could mean the difference between a job and the dole queue.

The average length of time sitting without a break rose from four minutes at day one to 15 minutes at two weeks. At the 12-month review, it was 43 minutes. Similar gains were made in the ability to stand. Then there was stair-climbing. On admission, folk could, on average, manage 71 stairs, a figure that more than doubled to 146 stairs at 12 months. The recruits also had to do a speed-walking test, to see how far they could walk in 12 minutes. The average went from 634 metres on day one to 1,034 metres at 12 months. And while all those gains were being made, there were big changes, too, in reliance on meds. When these people fronted on day one, their average dose of painkiller, measured as its equivalent in morphine, was 37 milligrams. At day 15, it had nosedived to just 0.6 milligrams. That figure rose again at 12 months, but to a still respectable 7.3 milligrams. Ratings of pain severity, depression, anxiety, and stress all dropped, as did the degree to which pain interfered with mood, sleep, relationships, work, and enjoyment of life. Catastrophic thoughts like 'it's terrible and I think it's never going to get any better' trended lower, too.

The numbers, in black and white, can only tell part of the story. But Trinca's team has corroboration in colour: a video is shot again at the three-week mark as their graduates re-run the gauntlet of Barbara Walker's long corridor. 'What is really fabulous is to see the difference with those videos, because they're walking faster, they look happier, they're stronger, and their posture's better,' Trinca says.

For Carl, who had endured two surgeries and years of pain, one part of the experience stands out. 'Without a doubt, the strongest effect would have been the pacing,' he says. 'I went from being able to sit for about ten minutes with no pain, to being able to sit for an hour with no pain. I wasn't really able to walk more than 100 metres without pain, but, not too long after, I could walk for a couple of kilometres. Now when I walk the dog, I can walk ten kilometres. I do still get the pain from walking, but it's at a level that I know I can deal with. I'm not exactly ever pain-free, but the pain is now something that I can manage.' Carl still takes tramadol, at much lower doses, but says the course was a turning point. 'It was without doubt the pivotal thing that got me back to being able to work full-time. No doubt about it.'

So can thinking change the nervous system? Carl is a straight-up kind of guy. He's not into reflecting on his thoughts or emotions, and confesses to believing that cognitive behaviour therapy (CBT) was 'mumbo jumbo' before he started the course. Now he uses elements of it — he'll do a 'body scan' in his mind, visualising his anatomy from top to toe in order to let go of stressed, racing thoughts. He also says learning how a sensitised nervous system can produce pain in the absence of injury was key: 'It all made perfect sense.'

Here is one way that thinking may change the nervous system. Understanding that moving is unlikely to cause injury when it is your hotwired nervous system pumping out pain may, as Australian physiotherapist Dave Moen writes in the title of his recent book, give people 'permission to move'. For Carl, that inexorable extension of daily movement did what glucose injections, radiofrequency ablations, stem-cell injections, and a shedload of pills couldn't. Somehow, it taught him not to be in pain.

The whole process, though, implies a concept that plenty of people might have trouble with — that Carl's pain, at least a good deal of it, was learned in the first place. But how on earth could you learn to hurt?

A khaki thread runs down from Crookwell in the New South Wales Southern Tablelands, past a mosaic of working farms and weekender plots, before striking north to enter valleys whose treed inclines are punctuated by granite bluffs that mark a path up the scarp face. This is the Wollondilly River, and, on one of those bush blocks, early on a fine May morning about 20 years back, Lorimer Moseley woke up, donned a sarong, and headed down to the river for a swim. The walk was uneventful apart from a momentary, twig-like scratch to the left leg that barely interrupted his stride. It wasn't long before Moseley stood at the river, which was at a low ebb during the big dry, tossed his sarong aside to greet the morning in all his glory, and waded in for a quick splash. The story ends there, because that's the last thing Moseley remembers. The scratch wasn't a twig but, rather, the bite of an eastern brown, the world's second-most-venomous snake.

The experience would temper many people's enthusiasm for bushwalking, but, not only did Moseley survive, his love for the Australian outdoors was undimmed. Not six months later, he was tramping in Lane Cove National Park, an idyllic bush oasis in Sydney's inner north where swamp wallabies forage among the wattle and scribbly gum, when he was stunned by a sudden pain in his left leg. It was so severe, he recalls, it threw him off the path. Here is how Moseley describes it in his book *Painful Yarns*:

> The pain shot up my leg like an electric bolt … I doubled over, fell backward onto a conveniently situated rock and gripped my leg. I couldn't stop my face from contorting and my eyes switching between looking for the snake like a madman for sanity, and then clamping tight. I was in agony.

When he had composed himself, Moseley looked down to inspect the area and found, this time, that he had indeed been scratched by a twig, which had left a tiny, solitary mark. So why would the bite of one of the world's most dangerous reptiles barely cause its victim to turn a hair, while a scratch from a twig threw him to the ground in agony?

Moseley answers that question in five short words: 'I think

it shows learning.' Moseley is not referring to the kind of learning you might do swotting for exams. He is talking about associative learning, which is when you come to understand, often unconsciously, that two things happen together. The most famous example of associative learning is Pavlov's dogs. They learned, during the course of the legendary physiologist's experiments, that the savoury delights of powdered meat were always announced with the ringing of a bell. At first, the animals only drooled when the meat came. But soon, they were salivating at the mere sound of the bell. This type of associative learning is called classical, or Pavlovian, conditioning.

Moseley thinks associative learning could explain his near-death twig experience in Lane Cove National Park. The snake bite at Wollondilly was almost painless. But everything tied to it — the scratch-like feeling, the body region of the leg, and walking in the scrabbly undergrowth of the Australian bush — was soon to be indelibly linked to a very close scrape with death. The subsequent twig incident embodied enough of those elements to connote danger, and so pain, the cardinal token that our wellbeing is imperilled, was triggered.

Moseley's theory is plausible, but proving it is another thing. For starters, you have to do an experiment that teaches people to be in pain. But how would you do that, and, even if you could, would anyone sign up? Both questions were front of mind for a young South African physiotherapist named Victoria Madden, who, in 2014, made the voyage from Cape Town to Adelaide to do her PhD with Moseley. Madden was on a mission to see if people really could learn pain by association. It just seemed to make sense. We link things, explains Madden, because it helps us to understand the world and protects us. For example, if you scald your hand in hot water as a child, you'll forever link hot water with danger, which is an excellent result. The body, however, has something of an agenda in this area — it tends to be overprotective. If the human body had a motto, it would probably be 'better safe than sorry'. This, says Madden, means the next time you rashly plunge your hand into a hot tub, you'll probably get pain when the water isn't quite so hot, all in the name of keeping you safe. 'If you feel pain sooner, you'll take your hand out sooner, and your skin won't be damaged,' she says.

Another problem is that we may, in fact, be *too* good at linking things. Madden tells the story of 'Debbie', a netball player who gives herself a nasty sprained ankle, scraping it on gravel during a match in icy weather. Oh, and a gent on the sidelines happens to be smoking a pungent cigar. As in a Sherlock Holmes adventure, all the details are relevant. 'Any of those pieces of information, the cigar smoke, a scrape against the ankle, the cold weather, could become cues of danger to Debbie,' says Madden. 'Even after her ankle ligament has healed completely, she might find that her ankle hurts a bit when she turns it inwards in cold weather, or when she smells cigar smoke, or when her ankle bone scrapes against the leg of her desk at work.'

Madden knew she wasn't alone in having a strong hunch that learned pain was a thing. It was something of an open secret in health circles that many patients had pain brought on by things that shouldn't be painful. Why not, thought Madden, get it out in the open? So she set about surveying a bunch of pain professionals, including doctors, physiotherapists, psychologists, and osteopaths. In the end, she heard back from more than a thousand pain practitioners across 57 countries. Madden asked them if pain could happen without a nociceptive, or typically painful, stimulus — 86 per cent said 'yes'. She asked if pain could be classically conditioned, like Debbie's netball injury, and 96 per cent said 'yes'. Finally, she asked if they thought there was scientific evidence that pain could be conditioned — 98 per cent said 'yes'. An overwhelming majority, it seemed, took a stance on pain that was remarkably enlightened, presumably based on their clinical experience. Except for one thing. On the question of hard scientific evidence for conditioned pain, they were downright wrong. Because there wasn't any such evidence.

Madden is blonde, fresh-faced, and speaks with a light South African accent in cadences that are crisp, upbeat, and lyrical. To address the knowledge gap, she also proved to be persistent and creative. She, Moseley, and a team of researchers spread the word they were on the hunt for study participants. These had to be adults who were pain-free and had never had chronic pain. Participants also had to have thick skin, literally. That's because they were going to get a laser shot at it — not the steel-cutting model, but the one dermatologists use to remove spider veins. Lasers use light and heat, and they can damage fragile skin. They also hurt, which is precisely what Madden wanted, in the nicest possible way. The other key piece of kit in her arsenal was the thing that makes your smartphone vibrate, a tactor.

Sixteen people took up the offer, and Madden and team got down to work. They had the group strip off their tops and lie on their tummies while listening to white noise through headphones — to mask the sound of the laser pedal being depressed. Then they began the pain training. First, they fired the laser at each person's back to find the lowest intensity that hurt. That done, they were ready to deliver a series of painful zaps, with a twist: a smartphone tactor was strapped just above the laser target zone on each person's back, and set up by Madden to vibrate as each zap was fired. In a third training session, the laser zap was turned down so it didn't hurt, and a tactor was strapped *just below* the laser target zone on each person's back — but the tactor was still set up to vibrate at the same time as each zap.

The volunteers were now ready for their pain exam. A laser calibrated to be just at the threshold of pain perception was zapped on their backs, paired with a tactor either above the spot or below it. What Madden and co wanted to know was this: did the position of the buzzing tactor influence whether the laser hurt? What they found would rocket the team into an elite cohort of scientists. When the laser was fired with the top buzzer, the volunteers felt it to be more intense, more unpleasant, and more often painful than zaps paired with the bottom buzzer. That's despite all the zaps having exactly the same intensity. What had happened? The upper buzzer had been linked to pain in the minds of volunteers. Now it could dupe them into thinking something that shouldn't hurt, did.

Madden had provided some of the first, if tentative, evidence that pain could be conditioned. That it could, quite simply, be learned. So how might this work in back pain? Let's say you bend over to lift a heavy box. As you straighten and twist with the box in your arms, you get a sharp pain in your back, an injury that lays you out in bed for several days. Through associative learning, the movement of bending forward, straightening, and twisting gets linked to pain in your mind's eye. That link, now etched on your brain cells, means that, when you bend and twist in future, it could trigger pain, even after the injury has healed. Bending over is like Madden's buzzer — it pushes the button on pain, even when you shouldn't have any.

But if Madden's study suggests pain can be classically conditioned, there is something important it doesn't tell us. Why does it only happen to some people? How is it that one person hurts their back and gets better with rest, painkillers, and a gradual return to movement, while someone else gets persistent pain that spreads?

One explanation harks right back to Pavlov's original, notoriously misrepresented experiments. The standard story, as I mentioned, is that the Nobel Prize–winning Russian conditioned his hounds to drool by ringing a bell just before they got the meat powder. In fact, Pavlov used a metronome, whose regular ticking worked a treat as an unconscious harbinger of meat. But there is another, even less well-known nugget buried in those early experiments. Pavlov noticed that, once the dogs learned to drool to the sound of a metronome, other noises could also get them slobbering, even if they were never paired with the meat powder. Bells and whistles, for example, got the canine juices flowing, too. The phenomenon is called stimulus generalisation because the original stimulus that brings on a behaviour — the metronome — gets generalised to similar things — bells and whistles — which are then empowered to bring on the same behaviour.

Moseley thinks the very same thing happens in pain, and he spelled out why in a groundbreaking paper written with Dutch researcher Johan Vlaeyen in 2015. If you hurt your back lifting a box, you may indeed learn to get pain from the exact same movement in future, way after things have healed. But you might also get pain from movements that merely resemble the original: tilting sideways, stretching backwards, and lifting your leg could also kick off pain. The first movement linked to pain, like Pavlov's metronome, gets generalised — not to bells and whistles, but to other movements. It's a theory that's supported by something very strange that happens to the bodies of people with chronic pain, something you can measure with one of the simplest tests imaginable — you just need a paper clip.

If you want to try it, open out the paper clip so the ends are about half a centimetre apart. Now press them into your fingertip. Do you feel two points, or just one? Most people will feel two points. But the palm of your hand is less sensitive — you need to open the paper clip to around a centimetre to be confident you feel two points. The back is even less sophisticated at the test, which is called two-point discrimination; the pins need to be around three to four centimetres apart before you can feel two prongs. But here's the thing: if you have chronic back pain, the distance gets even bigger. A review done in 2018 found that people with chronic lower-back pain needed the pins to be a full 12 millimetres wider to feel two spikes. So what's going on?

'What happens in chronic pain is that the brain's ability to precisely represent a body part becomes reduced,' says Moseley. Think of it like this. Your mind's eye, craning over to take a look at the clean curves of your back, is fitted with a pair of sand-blasted goggles that morph the geometry into flesh colours broken up with light and shade. Photorealism is replaced with the broad brush of impressionism. Moseley thinks this fuzzy representation of the back in the brain, evidenced by the two-prong test, is a sign that when the original painful movement was recorded in the brain, it too was fuzzy and ill-defined. All that fuzziness, he thinks, makes it far more likely that movements similar to the first, painful one, will hurt, too.

When Moseley first mentioned this to me, I found it a little esoteric, as you might. Why would the image of the back in the brain's usually crisp pictorial archive bleed into the page like a line drawing that thinks it's a watercolour? And why would chronic pain, following suit, spread to neighbouring regions? 'It's not that the pathology is spreading, it's that the protective system is increasing its area of protection,' explains Moseley. The brain, it seems, is doing its helicopter-parenting thing again. According to the theory, which Moseley calls the imprecision hypothesis, our noggins draw an iffy map of the injured area, then stake out an expanded zone with an alarmed fence to keep injurious movements out. The alarm signal, of course, is pain. The motive, you probably guessed, is 'better safe than sorry'.

All of which would be very interesting indeed if there was something we could do about it. So can we fit this newly myopic brain with a set of prescription specs to sharpen things up? And if we can, will it make the pain go away?

Moseley is pretty casual for an internationally famous pain researcher. When we chat, he's wearing a grey T-shirt under a dark open-necked jersey that would serve just as well on the sports pitch as in his busy research unit at the University of South Australia. He's 50ish, with receding dark hair kept extra short, expressive eyebrows, and clear, youthful skin. He is prudent about casting around catch-alls on pain — 'I'm no truth sheriff,' he advises me — but he has a quirky, quasi-spiritual side. When I first emailed him, I got a whimsical, poetic auto-reply: he would not see my message, it read, because he would be 'mindwandering' at 'bush and beach'. It might have been penned by 19th-century American transcendentalist Henry Thoreau. The overwhelming sense, however, is one of solidity and persistence. Moseley has had the chronic-pain bit between his teeth for a long time and is not about to let go. Like so many researchers, he's had personal experience of his subject — 'research is me-search', as the saying goes.

In 1988, a 17-year-old Moseley was practising a soccer manoeuvre made famous by the star player of the era, Diego Maradona. The bicycle kick is executed facing away from the goal — you jump to get airborne then fling the striking leg back over your head to fire the ball at the net. It is spectacular and athletic and needs compliant turf to cushion your back when you hit the ground. Unfortunately for Moseley, his landing was squarely onto the unyielding metal of a sprinkler head that had failed to retract into the grass. The injury to his back, he says, caused 'pretty rotten pain' for a decade. So Moseley had good reason, when he started his physiotherapy degree not long after, to pay special attention during the pain-science lectures. These, he recalls, included a decent slab on Clifford Woolf and central sensitisation.

'I would go into the biological science classes and think, "This is just the best, the human is incredible,"' he says. But the practical, treatment-based sessions were another matter entirely. 'I'd go into my clinical classes and think, "Do you guys not go to the bio classes?" Like, they didn't seem to relate, and the medical, physio-type advice that I got bore very little resemblance to my understanding of biological sciences.' It was a failure of translation that did not bode well for the young physio's nascent practice. 'I was working as a physiotherapist, treating people with persisting pain and just feeling useless. I thought, "I can't offer you anything." And no one had offered them anything that was helpful. And they were not believed, they were angry. I felt like saying to them, "Don't pay. You can't pay, because I got nothing."'

Yet unbeknownst to Moseley, beneficent forces were at work. Bound up in all that youthful candour, Moseley was unwittingly offering his patients something priceless. 'I would say, "This is why I think I can't help you," and I would explain my understanding of the biology of pain, and then notice that people would come back and say, "Ah, that was great. Can we just work more on that?" People would get better.' And what was he telling his patients that made such a difference? 'It was heavily informed by Clifford Woolf's central-sensitisation stuff. The main message was you're over-protected by a system that misunderstands the reality of the problem, and these are all the things that might be helping to drive your pain. These are the things you should be able to do about it — do you want to have a go?' Moseley has been on a mission to empower people with information ever since, carving out a niche as a global leader in pain research and education. He has produced a vast trove of studies, and his publications have been cited nearly 25,000 times. But, all the while, something has been bugging him.

One of the biggest problems for scientists is when they do a study and find that a treatment works. Wait, you say, surely they should be jumping for joy? Well, no. Imagine for a moment that you're the researcher and you've come up with a new treatment for back pain. For the sake of argument, let's say it's a six-week course of back massage. You enrol a hundred volunteers with back pain, put them through the course, and, at the end of six weeks, run a battery of tests to see how they did. Nine out of ten, you discover, are much better. That massage must be a wonder treatment, right? Not so fast. In fact, several things could have caused your stellar result. For a start, the natural history of most illnesses is to get better, even if you do nothing at all. Your volunteers' back pain might just have run its course. Then there is a statistical blip called 'regression towards the mean'. In healthcare, this means bad symptoms tend to revert, with time, to average severity, which generally means they improve.

And of course there's the famous placebo effect. People can get better, very simply, because they expect to. Placebo effects are especially powerful in conditions centred on the brain — like pain — where they can trigger an explosive release of endorphins, the body's inner painkillers. Placebo effects are also dizzyingly common, because just about anything can influence people's expectations of getting better. It might be the authority of the doctor's white coat, impressive plastic models of the spine sitting on their desk, or the upbeat way they introduce the treatment. I once interviewed a pain doctor who considered his tie to be part of the treatment, on the basis that it could relieve pain via a placebo effect. The upshot is a troubling one. If you happen to give someone a massage course while these other things are making the pain better, it's natural to think the massage did the trick. But it didn't.

All this means that pain researchers face a formidable challenge: how can they be sure it's actually their treatment making people better? This is where the nitty-gritty of study design comes into its own. What they have to do is give half their volunteers the real treatment and the other half a fake treatment. If the real treatment works best, then it's time to pop the champagne corks. But if the real treatment and the fake both work, then you know your real treatment is a dud. People are just getting better for one of those three reasons: natural history, regression to the mean, or the placebo effect. At this point, things get even stickier for back-pain researchers.

If your treatment is a drug, coming up with a fake drug is relatively easy. You just make a sugar pill that looks like the real one. This is your classic placebo — the word comes from the Latin 'I shall please', which the fake pill often does, when it works. But a big chunk of back-pain treatments are not drugs. Often, they are exercises, or psychological therapy like CBT, things that are exceedingly difficult to fake. So what do the researchers do? In effect, they give up trying to emulate the real treatment and go for something completely different. So the fake treatment might be an electrical nerve stimulator — a TENS machine — with its display lit up so it looks switched on, though it is secretly switched off. Or it could be acupuncture needles that are supposed to deliver an electrical impulse, but are not plugged in. The key thing is that people *believe* they are getting something that works, and so a placebo effect could theoretically happen. These are called sham treatments, and, Moseley tells me, they are a thorn in his side.

'In back-pain trials, the shams are nearly always crap. People know they're not getting a real treatment, and, in any pain study, this is a major problem for understanding what works, what was the mediating effect,' he says. But even when the group getting the sham believe it is real, there is a larger issue. 'I think so far any complex treatment that has a half-credible sham for back pain showed no effects.'

That is a big call, and depressing news if you've got back pain. It was also a devilish problem for Moseley because he, along with a small army of Australian pain researchers, led by James McAuley at Neuroscience Research Australia and Ben Wand at the University of Notre Dame, wanted to test their very own complex treatment. A good part of it would be based on the imprecision hypothesis, the brain's fogged-up image of the back. In 2015, the team secured funding for a trial called

RESOLVE, a name that would prove apt because they were up against it. To show the treatment worked, they would have to come up with a really good sham. Harder still, they would have to refocus the blurred brain goggles of a lot of people with back pain, or, as Moseley puts it, 'reinstate the normal precision of the brain's encoding of the back'. It was a tall order, so how would they do it?

The team started by enlisting 276 adults who'd put up with lower-back pain for at least 12 weeks. Half of them would be treated with the real McCoy, beginning with education based on Moseley's canonical book *Explain Pain*, co-written with Australian pain expert David Butler. The book, says Moseley, has a stand-out message: 'It aims to reconceptualise the problem as one of overprotection within your pain system, not pathology within your back.' Then the volunteers stepped up for 12 hours of training spread over 12–18 weeks. One of their tasks would be to 'redraw' their back upon the brain's very own canvas. Pushing the art analogy, if their painful back had the bleeding edges of an expressionist painting, the aim was to give it the hard-edged lines of an architectural drawing. But how, precisely, do you teach someone to recast an area of their body they almost never see, with the crystal clarity of a trained draughtsperson?

The researcher had each participant sit on an examination table, naked from the waist up, and then drew a grid of dots on their lower back — 16 in total, four up and four across — with a marker pen. Then the researcher handed their subject a picture of the grid with each dot numbered from one to 16. Which is when the poking and prodding began. Armed with a pencil-shaped tool, one end pointy and the other end fitted with a tiny, house-shaped eraser, the researcher jabbed every dot on the volunteer's back, one by one, with each end of the tool. The person had to answer two questions. First, was it the sharp or blunt end? Second, what number dot, based on the corresponding map they held in their hands, had the researcher pressed the pencil on?

A trend soon emerged. Most volunteers started off sluggishly, but, with each trial, they would get more answers right. Their brain's perception of the back was sharpening up. The researcher stepped it up and, taking a fine wooden stick, drew the letter 'A' on the person's back. 'What letter did I draw?' they asked. Then they traced the word 'CAT', the number '3', the sum '4 + 4', the volunteer all the while reporting, based purely on the feelings from their back, what was being written. This is called sensory retraining, and the responses were encouraging. But when it comes to getting sharp eyes on a sore back, there is, unfortunately, something else you need to do.

The brain doesn't just encode sensations coming from the back, what you might call the inbound route. It also has bits that execute the motor commands that produce movement. As if things couldn't get any worse for people with back pain, they have problems with this outbound route as well. In 2011, Moseley led a study that got people with back pain to look at a series of fairly repetitive photos — people twisting their torso to the left or right. The viewers only had to say which way the person in the photo was turned. That might sound easy, but Moseley's team made it harder by varying the photos — 56 in all — so that the model was twisted anywhere between five and 90 degrees. So how did the participants do? A group of healthy control subjects got the correct answer, as you might expect, 87 per cent of the time. But people with back pain were woeful. They got it right just 53 per cent of the time, little better than chance. So what was going on?

'As a participant, your brain is matching the posture in the picture. You immediately manoeuvre your own back, in your brain, to match the posture, and then you respond because you've worked out what the answer is,' says Moseley. This probably happens through mirror neurons, which are brain cells for an action that fire when you see someone doing the same thing. Mirror neurons explain how babies smile automatically when they see their mothers smile, minus any kind of practice. But mirroring back movements in the minds of people with back pain is wonky donkey. Why? A recent study hints at an answer.

It just so happens that two muscles, one called lumbar longissimus and the other deep multifidus, run down either side of your spine and help stabilise the back. Each of these muscles is usually controlled by separate bits of your brain's motor cortex. Usually, that is, unless you have chronic lower-back pain. Australian researchers used magnets to stimulate the brain's motor cortex and measured the response in each muscle. Normally, the brain bits that control each muscle are separate. In people with back pain, they weren't — they were overlapping. Those brain areas had become, as the authors put it, 'smudged'. Which looks like an elaborate way of stuffing up a very important job, so why would the body do this?

Again, its intentions are good. One result of a smudged back map in the brain is a loss of fine control over individual strap muscles of the back. That means back muscles tend to get activated en masse. This group effort is a plus, because the muscles stabilise the spine and make it harder to use, two things you definitely want after an injury. Bad things happen, however, if the muscles stay bunched up too long. Tight muscles do more harm than good, because you need a flexible spine to share load evenly across the discs — which act as the spine's shock absorbers — and among its supporting muscles and tendons. The back's biomechanics get thrown off, and that triggers pain.

In 2016, the RESOLVE team was ready to try and correct this blip. As in the 2011 study, they showed participants the pictures of people twisting one way or the other and asked them to say if it was to the left or right. Then they made it harder, gradually shortening the time people had to answer. Upping the ante further, they showed the participants videos of healthy people doing various back movements and lifting heavier and heavier loads — the participants were asked to watch like a hawk and imagine doing the movements themselves. The aim of this 'motor empathy training' was to sharpen the brain's motor control, leading to back movements with greater coordination, which could better handle load. Pain would, if the theory was on song, also get better. Then the participants had to do real movements with gradually increasing loads. They were given feedback to ensure their back movements became fluid, smooth, and, most importantly, precise, with movements of the lumbar spine, hips, and thoracic spine carefully separated. The whole program is formally known as 'sensory and motor retraining with graded motor imagery'.

Did it work? The researchers went to extraordinary lengths to answer this question, much of it focused on getting the sham condition to hit the sweet spot. The control group got a range of shams. Instead of pain education, they got 'verbal ventilation' — instruction on the anatomy of the back and an opportunity to air their pain experience with a therapist, but with no formal counselling. The sham for the motor training was another kettle of fish altogether. It was designed to mimic a therapy that could also alter brain wiring, or neuroplasticity. These patients got high-tech brain stimulators, clipped to the ear, that deliver a low-level electric current to the brain. But in this case, the devices came with an important difference. They were turned off.

Finally, at the end of 18 weeks, the results were in. The team was ecstatic. 'We identified significant improvement in people with chronic lower-back pain,' says Moseley. The study, published in *JAMA* in August 2022, found the 'real' treatment group had a full point reduction on a ten-point scale in pain severity compared to the sham control. 'Now that might not sound very exciting,' says Moseley, before making his point about the lack of back-pain treatments that show any benefit in the face of a credible sham. So what did the RESOLVE participants think of their 'brain stimulators'? 'We have tested the credibility of it, and the sham was perceived as more credible than actual treatment in this trial,' says Moseley. 'It actually does nothing, but some people got such great relief that they asked to buy them after the trial.' No, the team didn't sell any. But the success of the sham makes the RESOLVE findings seem encouraging. Treatments based on the imprecision hypothesis look to have real effects, superior even to a fake brain stimulator that worked so well that an enterprising salesperson could have shifted a ton out the door. So if you have back pain, should you sign up for a course of sensory and motor retraining? One factor in the decision, and there is no getting around it, is that the suite of treatments in RESOLVE is hard graft.

Carl wasn't in the study, but his recovery was also one of time and effort, which could well prove the adage of the Greek tragedian Sophocles that 'without labour, nothing prospers'. I asked Carl if, after his own gruelling labours, he had a message for other people in pain. 'It's character building,' he says. At first, I think he's being tongue-in-cheek, but he's deadly serious. 'No, it is, it's character building. It's going to change who you are, it should change who you are for the better. You just have to have an open mind to it. If you don't have an open mind, it's not going to work.' I delve a bit: 'So this has made you a better person?' Carl doesn't hesitate: 'Yep, it was just another journey down the rabbit hole of who I am, of self-discovery. It was just another major obstacle that I was able to overcome in my life.'

The revelation comes as a shock, because Carl hadn't been overly philosophical about his nine-year journey through pain thus far. But it's clear he had to dig deep, and the man that emerged at the end, while of the same parts, had a very different composition. All of which fits with an emerging concept called bioplasticity — the ability of nerves and other body structures to adapt and change to accommodate new learning.

Virtues of work notwithstanding, putting a fraudulent bit of kit on your ear does seem to deliver a lot of bang for your buck. The fake brain stimulator worked, quite simply, because people thought it would, a classic placebo effect. People were shifting their pain just by thinking differently. Which raises a provocative question: might there be a way to ditch the dud device and simply harness your very own thoughts to get rid of pain?

Chapter 3

An Old, Rusty Robot

how new beliefs can rewire your brain

At the dawn of a hot day in 1998, in an unremarkable factory on the outskirts of the Gold Coast, a sun-soaked Australian metropolis, Tyrone Cole turned up for his job as a boat builder. Then in his 20s, Cole was something of a tough guy, raised in one of the rougher suburbs of nearby Brisbane, where the street brawl had been a rite of passage. He'd managed to curb his flailing fists, however, with a martial-arts practice known as Chow Gar, rising through the ranks to win a swag of tournaments across Australia and New Zealand. Cole was a bona fide champion with a physique to match, lithe and hardened by discipline. But he was about to meet an adversary that would put his Chow Gar opponents in the shade.

'It was just a normal Gold Coast day where we get to work nice and early, 5.30, to work with the fibreglass while it's still cool,' Cole remembers. One of his jobs was to wield a chopper gun, which dices then fires strands of fibreglass into a wooden mould to create the deck of the boat. A chopper-gun rig is heavy. The gun is attached by a hose and metal arm to a 44-gallon drum of resin sitting on a trolley, which Cole had to pull with a big T-bar handle across the mould to get to where he was laying the fibreglass.

On this particular deck, there was a stubborn ridge he had to inch the rig across. It wouldn't go. Walking backwards, Cole pulled. It still wouldn't go. So Cole gave it the full force of his powerful frame, which is when the entire handle split, pistol-like, from the rig. 'As it snapped, I was fully lifting and leaning back, and my momentum just took off. I went flying backwards, and the steel handle went with me.' The horizontal kick threw Cole about five metres, but it was the vertical drop that did the damage. He fell a metre onto the hardened steel tubing of a boat trailer, his lower back striking the metal, his arms and legs flinging back like a ragdoll's. 'It felt like an explosion had gone off inside my back, and I thought I'd opened myself up. I didn't know how bad the injury was, but I couldn't move. I was in a real bad way.'

Cole was duly conveyed to the company doctor, who told him, 'Your spine's got a lump like half a cricket ball coming out of it,' and called timeout on work for three weeks. When his brief convalescence was up, Cole booked in to see another doctor to get cleared. 'In those days, I was really fit and I was very flexible, and he told me to touch my toes and I couldn't. I got just past my knees,' says Cole. 'Normally, I'd be touching the ground no problems. And that was it. That was my assessment.

He goes, "Yep, you're right, you go back to work."' Cole, now 47, laughs, with a tinge of bitterness. 'So he sent me back to work. But after three weeks, I still couldn't move around and do my job. I was unable to do the fibreglassing or anything. Yeah. So I lost my job building boats.'

In the years that followed, Cole's work ethic drove him as hard as his body would allow. He got a job as a firearms instructor. Then he found work at a shooting academy as a range officer, ensuring safety and compliance. That job lasted four years, but, periodically, his back would let him know it was there. 'I worked probably 60 to 80 hours a week. I was able to get through it, but, when I was tired, or if I'd done something wrong, lifted a box of ammo or twisted, then I was in strife.' If that happened, it usually meant an ambulance, a day in hospital, painkillers, and confinement to quarters for a week. 'But it was usually only once or twice a year.'

Brisbane is an hour's drive north from the Gold Coast. It sits on Moreton Bay and looks across to a wedge-shaped island that pokes like an echidna quill into the turquoise waters of the Coral Sea. This is North Stradbroke Island, or Straddie to the locals, and, since the mid-20th century, it has been mined for its precious mineral sands. Straddie silica is used to make glass for bottles, windscreens, and solar panels, and its zircon goes into ceramic tiles. All that digging has left its mark. The excavations have carved out gashes — inverted yellow dunes with grey spoil heaped alongside — amid the otherwise uniform greenery of tea-tree, eucalypt, and understorey shrubs that carpet the island. The damage is inevitable with dredge mining, the main technique used on Straddie. The miners dig a huge hole then fill it with water to make a slurry, which, replete with mineral bounty, gets slurped out through heavy-duty pipes to be processed.

In 2013, Cole landed a job in one of those dredge mines. His job was to dig trenches then lay the 40-foot steel slurry pipes. That year, four days after his 40th birthday, Cole had dug his trench and was walking backwards in the soft white sand, signalling to the guy driving the loader to lift the next pipe, when he stepped in his own ditch, rolled his ankle, and went down. 'I tore the ligaments off the outside of my ankle and destroyed a lot of the inside of the actual joint,' says Cole. 'And from there, it has been nothing but mayhem.'

The injury needed multiple surgeries to reattach ligaments and restore the anatomy with screws. 'I never got stability back in my ankle after the surgery. I couldn't weight-bear, because my ankle just kept collapsing on me. It took about two years before I could walk further than 100 metres. I was in a great deal of pain, so they pumped me up with all sorts of painkillers.' The company treated him well, keeping him on with light duties, but Cole could see the writing on the wall. His survival instincts kicked in and he started doing tickets — certifications to let him pivot to doable jobs in the mining industry. If there was a course on, he'd take it. He studied working at heights and in confined spaces. He started a course in mine management. He did 17 tickets in all. Pretty soon, however, those visions of an alternative future began to pixelate and distort, evidence of a coming outage to his connection to work.

Cole is stocky, with clear blue eyes, pale, unlined skin, and a smatter of greying stubble above a full chin. There's still a bit of badass about him. His gaze isn't quite a death stare, but it does have a certain 'don't try me' swagger. There's a hint of angry emoji, too, in his down-slanted eyebrows, and a tough-guy flourish in the worn peak of his black baseball cap, sunnies perched on top. But deep shadows under his eyes suggest vulnerability, and the bamboo pattern on his black shirt reminds me of the Eastern philosophy behind Chow Gar: to balance the opposing forces of Yin and Yang and be imbued with the vital life force of chi, the Chow Gar practitioner must look inwards. Cole's next words take their leave from this inner place, and, as they emerge, his voice falters and rises in pitch.

'They said they let me go because I had to park my car and walk a certain distance to the office. It was soft sand, and I was unable to walk on uneven ground. So I lost my job in the mining industry. And that's when I lost everything, my rental house, my car, my motorbikes.' Cole is a family man, married with two children, all of which magnified the losses. 'Financially, I was wrecked. I come from a very strong background of being able to support my family, and being unable to provide for them killed me. It really wrecked me. I worked hard so they could have a good life, and I lost all of that.'

Things started to spiral. Forced to walk with a stick, Cole was 'limping like a penguin'. The new biomechanics stressed his back, and the pain flared up. He couldn't train at his beloved Chow Gar, and he stacked on weight, adding 40 kilos to his once hardened physique. He applied for job after job and was knocked back every time — 'back injury' on your resume is never a good look. Unstable on his feet, Cole took several tumbles. His back got worse. 'The nerve pain was incredible. It felt like my spine had compressed and was touching a nerve 24/7. But up to 30 times a day, I'd have a spasm if I bent down or if I moved funny. It would be like you plugged me into a wall and gave me an electric shock for what felt like 30 seconds at a time. I was so crippled I couldn't move. I couldn't drive. I couldn't push my brakes,' says Cole. 'They gave me Lyrica. Anything to do with painkillers and nerves, they gave it to me and I was on it, and nothing they gave me would take away any of the pain. All it was doing was drugging and messing me up.' Body and mind have their limits, and Cole was close to his.

'I was lost, and, to be honest with you, all the painkillers that I was on just accelerated my depression. It put me in a very bad, dark place. I was obsessed for probably two years with not being here anymore. Because I couldn't provide for my family.'

It was around this time that Cole met a man named Daniel Harvie. Harvie is a physiotherapist and pain researcher based at the leafy Gold Coast campus of Griffith University. Harvie is a calm, softly spoken man in his 30s, with a dark beard and wispy hair drawn back in a ponytail. When we chat, the weather is chill, and Harvie is rugged up in a thick Navajo cardigan over a grandad-collar T-shirt. Squint your eyes and you get visions of TV's medieval wizard Catweazle, which, notwithstanding the scientific rigour of Harvie's work, is apt, because Harvie has something of a preoccupation with the illusory.

One of Harvie's gigs is driving a bus, but not your run-of-the-mill jalopy. Once a year, he joins the Pain Revolution outreach tour as captain of the 'brain bus', a converted campervan with the image of a sprinting skeleton plastered on its side. Billed as a 'mobile illusion van', it's kitted out with everything needed to bamboozle brains in pain, from a virtual-reality system to a computer loaded with imagery that, to the novice viewer, defies rational explanation. I ask Harvie for a demo, and he obligingly loads up a PowerPoint.

An image appears on the computer screen. It's called the Cornsweet illusion. Two lozenge-shaped blocks are balanced precariously, one on top of the other. They are joined at a hinge and open out onto a field of what looks like 1950s Axminster carpet, its floral pattern receding into the distance under a dazzling blue sky.

The top block is black and the bottom one, quite obviously, white. Which is why Harvie's question comes a little out of the blue. 'Which of those two blocks looks darker to you?' he asks me. 'Well,' I say, a little hesitantly, 'the top block looks darker.' Now, with the hackneyed air of a conjurer who's done his trick too many times, Harvie tells me that, in fact, the blocks are exactly the same colour. To prove it, he puts a digital screen over the image, a big black square that shuts out the carpet and the sky completely, with two little windows that only let you see the blocks. The effect is dramatic. The blocks are clearly both grey. But how could black and white, the definition of polar opposites, be the same? It is, says Harvie, all about the sun.

'Basically, your brain is asking a question, and the question is something like "What colour is this?" To make that decision, it is considering information that is emanating directly from those locations. But it is also taking into account other information — for example, where the light is coming from.' Light in the image seems to come from above, striking the top block first and leaving the bottom block in shade. In the real world, dark things in direct sun look lighter. White things in the shade look darker. Your brain adjusts for this. The two blocks are grey, but your brain says the block on top in the light is probably darker — black — and the block on the bottom in the shade is probably lighter — white. You conclude the blocks are different, and get it wrong. Which is neat, but what's it got to do with pain?

'When it comes to pain, the brain is also asking a question, and that question is something like "Is there danger in the body?"' says Harvie. 'Of course, the brain is really interested in information that's emanating from the body part of interest. But it's also interested in all kinds of other information.' That information could include, says Harvie, something scary a doctor, physio, or friend might have told you, something worrying you read on an X-ray report, or whether a movement you just did was one that hurt in the past. And if that painful movement happened at work, even your workplace could make it into the brain's danger file. 'So, a bit like the case of the colour of the blocks, your brain's going to takes a best guess about whether you need to be protected.' The point is that, just like the brain thinks the blocks are black and white when they are both grey, it can generate pain when the body part has healed. It is all about context — the slant of the sun shifts colour perception, and the slant of your information shifts danger perception. The result can be pain with no danger, clear or present.

It's a tidy theory, and about as far from Descartes' single-circuit alarm bell as you can get. But let's face it, pain is often a screaming siren in your ear like an ambulance going flat stick. Pain is urgent; it comes unbidden. Could it really be a considered, albeit subconscious, review of some intelligence that your brain just ran its eagle eye across? Well, think about it for a minute. If pain was a pure sensation, untainted by any of the baggage of our thoughts and feelings, you'd expect it to converge on a neat little walnut of brain marked 'ouch'. Logically, that pain walnut would be in the sensory cortex, the part of the brain where sensations are registered. But when you scan the brains of people in pain, that is not what you see.

'When we photograph, using fMRI [functional MRI], the brains of people in pain, it is not just the sensory cortex that lights up,' says Harvie. 'It's a dozen or so areas. So I think our best read of this is that the brain is, in every pain scenario, integrating information from all over the brain. From information that might have originated from the body, or from vision, or from hearing, or from memory areas, or from fear centres. It is the brain concluding that the body is in danger, not on the basis of signals in one neural pathway, but on the basis of all the information that the brain has available to it.' This, says Harvie, is the neuromatrix (i.e. originating from a network of neurons in the brain) theory of pain, an idea first floated by Ronald Melzack in 1990 that has gained traction as brain-imaging tech has evolved over the subsequent three decades.

So, if the brain files a bunch of data on thoughts, feelings, and sensations to make a danger calculation that produces pain, and if the brain often gets the calculation wrong, what Harvie wanted to know was this: can you hack into the brain's intelligence-gathering exercise to get it back on track? Is it possible, he wondered, to give the brain different information to make the pain go away?

Harvie doesn't come across as a tech guru. He's brewed his own kombucha, and, as a musician, he favours a vintage 1969 Fender slimline guitar. He once wrote a song called 'Homesick' about the travails of drought-stricken farmers — the music video shows him sticking out a hopeful thumb to hitchhike on a lonely, outback, red-dirt road. The flavour is organic and homespun. But Harvie's approach to pain management would be bleeding edge. He would leverage, he decided, the awesome, immersive power of virtual reality.

Working in a team that included Lorimer Moseley and Victoria Madden, Harvie scouted for volunteers. They had to have chronic neck pain and be unafraid to turn heads in the virtual world — their own, that is. Harvie signed up 18 women and six men. They'd had neck pain for an average of 11 years, from whiplash, degenerative spinal disease, and repeated strain. One by one, Harvie ran them through the experiment. Each person sat in a chair with a belt strapped around their shoulders to stop them twisting. Then they put on headphones with white noise to drown out the real world. Finally, they strapped on a set of virtual-reality goggles. It was time to enter a new cyber reality, which could be a park, the bucolic countryside, the grounds of a church, or somebody's dining room. Harvie now issued two simple instructions: 'Turn your head slowly to the left and stop when it hurts,' and, 'Turn your head slowly to the right and stop when it hurts.' The directives were basic, but the VR setup was anything but. Harvie was measuring exactly how far each person turned their head, with motion sensors fitted in the goggles. The measurement was critical because, even inside the new order of this manufactured world, all was not as it seemed.

Harvie was secretly manipulating the virtual scene. Sometimes, the countryside would sweep past a little further than it should, so the person thought they had turned their head 20 per cent more than they actually had. Sometimes, Harvie would make the countryside go by at a more stately pace. Now the person thought they'd turned their head 20 per cent less than they had. For comparison, sometimes the cyber world turned at exactly the same rate as the real world. What Harvie wanted to know was, quite simply, if distorted information in the virtual world could affect pain in the real world. The results were stunning.

'If I made it look to them like they were turning more than they were, then their pain came on sooner in the movement. And if I made it look like less, their pain would come on later,' says Harvie. Let's put hard numbers on that. When the person thought they had turned their head further in the virtual park, countryside, church, or dining room, pain came on at a neck rotation six degrees earlier than it normally would. Incredibly, when people thought they had turned their head less in cyberspace, they could turn their heads seven degrees further before getting any pain. For Harvie, the conclusion is about as concrete as you can get. 'It showed that their pain wasn't only dependent on the actual physical stresses. It depended on how far their brain thought they were turning.'

So pain is more like an inference, a conclusion about danger based on all available evidence. And inferences are only as good as the evidence on which they are based. But just what kind of dubious evidence was the body sending to the brains of people in pain? Harvie started searching, raking over some choice coals in the brazier of pain theory. Among the embers, he found an experiment that could be key to correcting the brain's pain mistake. It's called the rubber-hand illusion.

To understand the rubber-hand illusion, it's well worth giving it a try. If you're up for it, find a cardboard box big enough to put your hand and forearm inside, cut a semicircle from one end of the box, just big enough for your forearm, and remove the other end of the box completely. Then sit in a chair with both hands resting on a table. Have a helper insert you right forearm into the box through the semicircle. Take a rubber glove and place it just to the left of the box, in plain sight where your right hand could comfortably be. Now drape a towel over the cuff of the glove and your right arm — your right hand is in the box, so you can't see it, just the rubber glove poking out from the towel. At this point, your helper will need two small paintbrushes. Have them stroke the index fingers of the rubber glove and — through the open end of the box — your hidden right hand at the same time. Where do you feel the stroking? If the illusion works, which it does a lot, you'll feel the index finger of the *rubber hand* being stroked. It is a bizarre sensation. Scientists call it an 'illusion of body ownership', because we get duped into thinking something is part of our body when it isn't. Lawrence Rosenblum, a psychology professor at the University of California, often adds a second part to the trick. When subjects say they can feel the brush on the rubber finger, he gets out a hammer and bashes it. If you had any doubt that people adopt the glove as their own, it is dispelled when you see them freak out.

What does this supreme party trick have to do with treating pain? When I was a junior doctor at Prince Henry's Hospital in Melbourne in the 1980s, the vascular surgery ward was populated by folk with a very common problem. They had poor circulation to their legs, often from diabetes or smoking, and had developed gangrene of the foot. In many cases, the leg had to be amputated below the knee. Afterwards, some of those people were left with the strangest of feelings. The lost leg felt like it was still there, sometimes itchy, often painful, almost always distressing. This is the well-known phantom-limb syndrome, and, on top of the disability of being an amputee, it causes immense suffering. Someone who had lost an arm after a motorbike accident described it this way: 'It's as if the skin of my arm has been ripped off, salt is being poured on it, and then it's thrust into fire.' And it's here that our malleable, plastic sense of our own body, exemplified by the rubber-hand illusion, can come to the rescue.

'Sometimes people's phantoms are in a disfigured position,' says Harvie. 'Say their hand before it was amputated was crushed. Sometimes they feel that their hand is still in that crushed position.' In such cases, therapists have a low-tech fix that can yield big rewards. They sit the person down and have them rest the remaining hand on a tabletop. In the spot where the missing hand would have been, the therapist sets up a vertical mirror. In a magical flourish, the patient now sees their amputated hand reappear — as the intact, mirror image of the existing hand. Numerous studies suggest this can help phantom-limb pain. How? 'Pain might, in part, emerge from the brain holding a representation of that body part as being injured and in need of protection,' says Harvie. 'From mirror therapy, they can use the reflected limb to sort of unravel that phantom into a more normal posture. So that might, in effect, be reversing that unhealthy, injured representation of the body part.' It is a very cool body hack. Convince the brain that the injured part is fixed and it dials down the danger detector — pain. But Harvie's patients were not amputees. What could he do to help, say, back-pain patients?

When it comes to illusions of body ownership, Harvie knew virtual reality left the rubber hand in the dust. Our brains find the virtual world irresistibly real. One macabre study had participants in VR goggles rest their hand on a desktop, which was replicated in the virtual world. In perhaps the first ever case of cyber sadism, the experimenter thrust a virtual knife into the participant's virtual hand and virtual blood spilled everywhere. Not only were the volunteers scared that they might be hurt by the knife, but recordings of their brainwaves showed that the motor cortex was primed to pull the corresponding hand away, just as it would in real life. Harvie wanted to tap into this verisimilitude.

In 2018, Tyrone Cole was going to a pain-management course at Gold Coast University Hospital. It had helped him cut down his pain medication and get moving again, but he was still at a low ebb, feeling like an 'overweight cripple'. Then a senior physio on the course mentioned that a man called Daniel Harvie wanted to use VR to treat pain. So Cole hauled himself over to Harvie's office at Griffith Uni.

'I went into Dan's thinking, "Nothing's worked for me so far, so I'll give anything a crack,"' says Cole. 'Dan put the VR on me, and he goes, "Let's try boxing, you've done that before."' The game Harvie chose wasn't Chow Gar, but it would have to do. It was called *Creed: Rise to Glory*. You enter the body of Adonis Creed, a wall-punching, sinew-straining prize fighter and star of the Creed franchise of movies, which brings Sylvester Stallone's Rocky story into the 21st century. In the first movie, released in

2015, Creed sets out his philosophy in plain language: 'I been fighting my whole life. It's not a choice for me.' The same had been true for Cole, but, right now, that's where the similarities ended.

'For the last five years, I couldn't throw punches. I could do the motion, but I couldn't put effort into it without hurting my back,' says Cole. 'If I just stood there and went box, box, threw two punches, it would twinge my spine so badly. All I've done, all my life, was train martial arts and punching. But I couldn't do it. A 12-year-old would have outdone me,' he says, with a laugh.

But in Harvie's lab, with Cole's hands gripping the controllers that became his gloved fists in the virtual boxing ring, where he faced an opponent that looked like Mike Tyson's bigger, badder brother, something shifted. 'I was slowly rotating, little bit by little bit. I was doing the punches and the motions of boxing, and, pretty rapidly, the first time I tried it, I realised I was able to move a lot more,' Cole recalls. 'It was as if it was lubricating me up, an old, rusty robot. I was able to move a lot more than what I could even contemplate for the last five years. It was working. It blew me away.' Harvie gave Cole a VR system to use at home. 'All of a sudden, I was able to put a little bit of power in the hips. It was just nice to be standing in my living room with the headset on, getting in the zone of having an opponent. I'm working the body, and, every now and then, tap, tap, there's more weight, more effort I can put into my hips.' It is unlikely that Metro-Goldwyn-Mayer Pictures, one of the companies behind the first Creed movie, envisaged their hero as a therapeutic tool for people crippled with back pain. But entering the Adonis avatar was part of an alchemical transformation for Cole.

Harvie wondered if a switch from the formulaic moves of the fighting arena could push things along further. A VR game called *The Climb* features a number of high-altitude venues, one of which is a badlands-style rockscape of sheer horsts and pinnacles broken up by plunging canyons. The climber negotiates vertical rock ladders and manoeuvres sideways with just a hand on a shallow ledge preventing a fatal drop to the valley below. I watched a video of an experienced climber doing it — it's physical and scary, so much so that he flinched when he fell. How would Cole handle it?

'Everything I tried before that was pushing away from my body. But as I was rock climbing, I realised I'm actually grabbing, pulling up, and pulling in to my body. And I could never do that at all. I realised that, shit, I haven't been able to do that movement for five years, and here I am actually doing it.'

Moving in VR is, of course, very different from the stresses of the real world. You're not suspending your body weight off those vertical cliffs. Your driving punches aren't meeting the fleshy resistance of an opponent's body. Could Cole take hold of his new mobility and hoist it into reality? 'I had so much nerve pain in my spine that I literally couldn't do things. But from the virtual reality, I'd walk away, I'd take the goggles off, and then, all of a sudden, I could do it without the goggles,' says Cole. The dismal state of his finances, however, meant he was going to have to get creative. 'I couldn't afford the virtual reality. So I bought a punching bag. I bought a physio balance board.

And I started training without the virtual reality.' After all those virtual wanderings, Cole finally ended up in a new place.

'Now I don't take any painkillers whatsoever. I'm not saying I live a life without pain, because, every now and then, it tells me I've got pain, that's for sure. But I don't take any painkillers. I'm not on any antidepressants. I'm not on anything, because now, I suppose, I've re-educated myself.'

Harvie documented Cole's amazing results in a paper published in 2020 in the journal *Frontiers in Virtual Reality*. Cole notched up a number of changes that are mighty impressive. The therapy happened over four weeks and included nine 15-minute sessions, three face-to-face and six at home. On a raft of body-image measures, including how strong and agile Cole felt, and how confident he was to do everyday activities, he gained between three and seven points out of ten. And on questions of self-efficacy — whether Cole believed he could live a normal life and achieve most of his goals despite the pain — he shot up from two to ten out of a possible 12 points. And the all-important question of pain? It dropped from a five to a three out of ten. Harvie, as a rigorous scientist, is cautious about labelling VR a wonder treatment for pain, and Cole's experience is a single case study, not a controlled trial — there are none — so the usual caveats apply. Cole could have got better due to the natural history of the pain process, or his symptoms might have regressed to the mean. Perhaps virtual reality produced a placebo effect. Still, Harvie conjectures that entering the pumped, fit-for-purpose body of Adonis Creed may well have updated Cole's mistaken, still-injured body image.

If VR was responsible, it is something of a paradox. It was an illusion that led Cole to update his perceptions and beliefs about his body and what it could do so that they were more in line with reality. Cole's revised body image had less need of protection, and so his brain made less pain. But does it take an elaborate therapy program and a virtual-reality kit to shift your beliefs?

The short answer is no, it doesn't.

About 20 years back, Lorimer Moseley was working in his physiotherapy clinic when a patient came in. She was middle-aged but hardly making the most of life. The woman was confined to a wheelchair thanks to 25 years of overwhelming back pain. Moseley got down to work. He took a history, delving deep into the trajectory of her pain and the things that made it better or worse. He scoured the tests and scans she'd had over the years. Then he did a rigorous examination, pushing and prodding her back, asking her to lift her legs, and checking her sensation and reflexes. At the end of the evaluation, Moseley was left with a single, inescapable conclusion: there was simply no evidence of an injury warranting such severe disability.

In all, Moseley spent three hours with her, concluding with an explanation of why he thought the soft tissues of her back were okay and checking that she understood. Satisfied she did, Moseley said goodbye and booked her in for review, hopeful his reassurances might bear fruit. 'She came back a week later, walked in, looked ten years younger, looked a million bucks,' says Moseley. 'I said to her, "So why do you think you've improved?"

waiting for her to sing my praises. She said, "Well, after I left you, I finally had the appointment with my sister's clairvoyant, and, as I was leaving, she said you can get out of that wheelchair because there's nothing wrong in your back anymore. And I woke up the next day and my back hasn't hurt since."' Like the professional he is, Moseley swallowed his wounded pride, accepted that his dedicated explanation of pain science had come second fiddle to a fortune teller, and asked the woman if she'd mind if they followed her up.

Moseley's team reviewed her regularly over the next 11 months. She stayed pain-free, ditched the wheelchair, and got back to full-time work. How does Moseley explain the woman arising, dramatic and Lazarus-like, from the chair that was her prison for so many years? 'It makes me think, if the brain's convinced by anything that you don't need protection, it just won't make pain.'

Katja Wiech is a psychologist and neuroscientist at the University of Oxford whose entire professional life has been dedicated to finding out how beliefs affect the perception of pain. 'A very popular belief amongst chronic back-pain patients is that there is something degenerative in their spine and that causes the pain, which obviously fuels their anxiety,' says Wiech. 'But even though they very often undergo diagnostic imaging and nothing can be found, they just hold on to the idea because it seems to match their perception. What they feel seems to be best explained by the idea that something terrible is happening in their body.' Wiech's mission is to dissect those beliefs, see how they drive pain, and, ultimately, discover if there is any hope of changing them.

It's midmorning in Oxford, and Wiech is fresh-faced, smiling, and relaxed, with her reading glasses pushed back as a head band and her shirt collar upturned over a dark V-neck sweater. Her animation is compelling, but I can't help but glance over her left shoulder. There, resting on a white tallboy, is a stunning portrait of a young woman, eyebrows raised haughtily, in the garb of medieval aristocracy. She wears a three-chain necklace, and her hair disappears under a conical hennin, the fashionable headdress of the day. It is striking, not just aesthetically, but because Wiech uses the female portrait as an investigative tool.

A little over a decade ago Wiech teamed up with a psychologist called Miguel Farias, whose special interest was religion and spirituality. The pair had a mutual fascination with one of the most famous, and mystifying, figures of the last millennium. Joan of Arc's military deeds played no small part in elevating her to the position of patron saint of France. But it was another of the heroine's traits that piqued the researchers' curiosity. In 1429, during the siege of Orléans, at the latter end of the Hundred Years War between France and England, Joan reportedly took an arrow to the left armpit. Story has it the arrow penetrated 15 centimetres, but Joan, in a feat of superhuman stoicism, carried on with battle. She was later captured by the English and burned at the stake, barely uttering a word as she was consumed by flames. Six hundred years later, some of the facts are sketchy; that Joan was a pious Catholic, however, seems certain, and many have suggested her faith helped her endure what must have been unspeakable pain. Now Wiech and Farias wanted to settle the question. Could religious belief, they wondered, really make things hurt less, and, if so, how?

The pair enlisted the help of 12 young Catholics from Oxford who were regular churchgoers, prayed daily, and attended the confessional on an as-needed basis. The group was joined by a dozen atheists who affirmed they had neither religious nor spiritual inclinations. One by one, they entered an MRI scanner, with an electrode attached to the back of their left hand that would deliver a mild electric shock. Their task was simple. They had to rate how intense the shock was on a scale of zero to 100. Yet Wiech and Farias had added a twist, which is where the old pictures of young women came in. Shortly before half of the shocks, they showed all volunteers a picture of the Virgin Mary — *La Vergine Annunziata* by the Italian Baroque painter Sassoferrato. Before the remaining shocks, all the volunteers saw a similar, secular image, *Lady with an Ermine* by Leonardo da Vinci. The participants had to focus intently on the face of each woman while they got the shocks. What Wiech and Farias wanted to know was this: did Catholics looking at the mother of Jesus feel less pain?

They did. Catholics felt 12 per cent less pain looking at the Virgin Mary compared to gazing at *Lady with an Ermine*. The atheists, by contrast, felt the shocks to be equally intense whichever picture they looked at. An unkind interpretation of the study might see it as divine retribution for the faithless, but what the participants said at a debriefing suggests something altogether more intriguing. When Catholics and atheists viewed the Leonardo picture, they had similar responses — 'I liked the picture and found it interesting', 'The picture was purely aesthetic', 'She looked serene, chilled out', among them. But when the Catholics looked at the religious icon, their reports could not have been more different from those of the atheists. 'I felt calmed down and peaceful', 'I felt being taken care of', and 'I felt safe', they said. Those responses, Wiech and Farias concluded, suggest Catholics were using the Virgin Mary to reappraise the meaning of pain. In effect, they were reframing the experience from one that was threatening to one in which they felt safe.

Which leads to the second finding. As the Catholics basked in their superior pain relief, a brain region lit up on the MRI scan. It's called the ventrolateral prefrontal cortex (VLPFC), and it remained steadfastly silent in the atheists. What does this brain region do? First, it's known to be active when we rejig our understanding of things in order to feel better — for example, when we remind ourselves how many people are worse off than we are. But the VLPFC is also a key area in one of the operations of Melzack and Wall's spinal gate. You might remember the soldiers from the Anzio beachhead who, knowing they would soon leave the battlefield, felt no pain from their wounds. The VLPFC is a key relay whereby thoughts, feelings, and expectations can modify nerve activity in the spinal cord to tamp down pain signals.

So what does the Virgin Mary study tell us? Should doctors carry a miniature quattrocento portrait in their white coat pocket to ease the pain of their patients of faith? Maybe not. What the study does offer is a solid proof of principle — that putting a different spin on the meaning of pain can reduce it, probably by switching on nerve pathways that shut the spinal gate.

On the face it, that might sound easy, but, as anyone with persistent pain knows, thinking pain away is no cinch. The idea that your back is injured, and will hurt if you move, seems to lodge in the mind like a mammoth in tar. Another part of Wiech's research helps explain why. But it might be easier to understand I if give you an example. The example is that you probably read that last sentence wrong. Go back and read it again. Did you read it as 'if I' instead of 'I if'? If you did, don't worry — you're not alone. You read what you expected to see, something that reflects a fascinating thing our brain does when it tries to learn stuff.

Nearly all the sentences you read in the past had 'if I' as the correct order. As the years went by and the number of sentences you read piled higher, your expectations about how a sentence should look got cemented in, like a foundation stone at the bottom of a historic building. Now you just expect to see 'if I'. When your brain suddenly sees the reverse — 'I if' — it has two options. It can decide that 'I if' is right and it's time to update your woeful English. Or it can call it a typo and ignore it. Most of the time, your brain ignores it, quite unconsciously. This is called confirmation bias — the tendency to perceive things in ways that confirm our existing views. In this case, that 'if I' is usually correct. By and large, that's helpful, because we don't waste precious energy deciphering outliers such as 'I if'.

The 'I if' conundrum is, however, just a tiny skirmish in our titanic struggle to understand how the world really is. The bricks and mortar of our expectations are constantly being peppered with spitballs of new information. If the spitball fits our expectations, no action is needed. But if the new information deviates from what we expect, we have two options. We can update our expectations. Or we can ditch the new info. Which is precisely what happens to some people with pain. If you think about it, when you bend over, you also have two options. You can feel pain in your back, or not. If you live with persistent pain, you are going to expect it to hurt. For many such people, the 'no pain' option is like the 'I if' word combo: it is an error that gets discarded, an information spitball that, lacking the requisite tenacity, simply slides off the monumental edifice of your pain expectations.

'When you have back pain and you think a certain movement is going to hurt and you bend over, physiologically that wouldn't trigger any back pain,' explains Wiech. 'But because you have that strong expectation that it will, it starts hurting even before you get to the point where it would normally hurt.' There is a pitched battle between the pain you expect to feel and news that your back has healed. Oftentimes, the winner is clear. 'The tug of war would be won by expectation because it pulls along incoming information, and sort of amplifies the pain.' This is called predictive processing, because the brain processes information based on its own predictions — your own expectations.

Can you change your expectations? Not long after the Virgin Mary study, Wiech was part of a team that set out to try. They enlisted the help of 22 healthy people. The team was going to give each volunteer something bad and something good. The bad thing was a small, square device called a thermode, which they strapped to the outside of each person's calf. The thermode, as its name suggests, heats up and stings your skin with a precisely calibrated quantum of pain. The researchers cranked up the thermode until the volunteers rated the pain as seven out of ten — none too pleasant. The good, or relatively good, thing was a powerful opioid painkiller called remifentanil. All the volunteers got a continuous infusion of remifentanil via a drip in their arm. But everybody got something else as well. They got information, which, depending on where you sat, could be good, bad, or indifferent.

The researchers began the experiment by turning the thermode on and off to deliver the mini-burn, over and over. Then they turned on the drip, and the painkiller began trickling through each person's veins. Which is when the psychological operations began. The painkiller ran for a full 30 minutes without anyone telling the volunteers. How was their pain during this period? It dropped a little, evidence the drug was working. Then one of the researchers announced that the painkiller would now be 'started by the anaesthetist'. What do you think happened to the volunteers' pain? It got dramatically better, dropping by half, even though the painkiller was still running into their bodies at exactly the same dose. Merely believing the drug had started reduced their pain profoundly. Another ten minutes or so elapsed and the researcher piped up with another announcement. The painkiller, they said, 'would now be stopped'. Guess what happened to the volunteers' pain? It bounced back, and not just a little. Their pain was now as bad as it was at the beginning, before the remifentanil had even started. But the drug was still running in, unchanged.

Let's take a moment to consider what that means. Remifentanil is no ordinary painkiller. It's used when you give someone a general anaesthetic, often for major operations like brain surgery, a heart bypass, or a spinal fusion. When it comes to painkillers, remifentanil is major. But the mere act of thinking it had been stopped completely wiped out any of remifentanil's effects. Are you getting an idea of just how powerful expectations can be? Expectations are king, dictating pain so completely that not even a supreme painkiller can fully counter their effects. The team was also running brain scans on the volunteers. You might not be surprised to learn which parts of the brain made their presence felt — the ones that form part of the descending pain pathway. The ones that close the spinal gate.

It is, of course, the amazing placebo effect, spelled out in bold. We dial pain up and down based on what we expect because, as this group of healthy volunteers attested, we are all wired that way. Does that make the pain any less real? Not at all — the volunteers' pain rankings were plain for anyone to see. Pain is real. It just may not mean what you think it does.

I asked Katja Wiech about Lorimer Moseley's patient. Could her stunning exit from the wheelchair have come about because the clairvoyant was able to shift her expectations? 'Absolutely. You do change your expectations, so incoming information isn't met with an amplifying force, but with something that puts it into place,' says Wiech.

In each case — the woman who rose from a wheelchair by clairvoyant's decree, the Catholics who bore shocks better when contemplating the Virgin Mary, and the remifentanil recipients primed to think differently by the presence of an anaesthetist — expectations were manipulated to alter pain. People reappraised their situation, and it made their pain better. But if you want to use these examples to treat pain, you run into trouble. Many people don't care for religion or clairvoyants. Many more are reluctant to rely on well-intentioned falsehoods from doctors, even if it means pain relief. So what about true beliefs — can they bring about meaningful pain relief?

Tim Salomons has been spending a bit of time at home recently. A psychologist and neuroscientist, Salomons divides his time between the University of Reading in the United Kingdom and Queens University in Canada. His research has, this year, been 'Covided', as he puts it, the pandemic punching a virus-shaped hole in his research schedule and putting a limit on his transatlantic gallivanting. Home is Kingston, Ontario, where Salomons' office has something of a shamanic cave feel to it. To one side, an appliqué wall hanging renders a New Age scene of rocks and a mountain with the streaming rays of a stylised sun rising behind. Enigmatic paintings of a bleak industrial landscape and indiscernible words bleeding down a wall lean on a shelf behind his desk. Get up close and you realise these are, in fact, album covers and that Salomons has a penchant for vinyl, preferably recorded by raspy-voiced Englishmen in the 1970s. It is music that melds poetry with a deep concern for the human condition, something not too far from Salomons' own raison d'être.

'What has sustained me is hearing from pain patients how frequently they are the lost population of the medical system,' says Salomons. 'They're suffering every day, and people treat them like they have some sort of nutty cuckoo disease. They imply that it's fake.' There is a phrase that many pain patients fear above all others: 'It's all in your head.' This is the bogeyman, a throwaway line that, in a single swoop, dismisses pain as something unreal. But, as all the foregoing shows, pain can be both real *and* in our heads — the head holds the brain, and the brain makes pain. Salomons, however, had a hunch that the brain was doing something else, that it had more of a role in pain than even the most-enlightened researchers had thought possible. What if thoughts and feelings, he wondered, were causing physical changes distant from the brain, in places where pain is processed, like the spinal cord? Salomons, working with a crack team of scientists at the University of Toronto, was determined to find out.

They issued a call for volunteers on the university campus and among workers at the nearby Toronto Western Hospital, a series of brown brick monoliths near the city's downtown area. The job description had a familiar ring: they were seeking healthy people with no history of chronic pain who were willing to be hurt — under controlled conditions, naturally. In the end, 34 people in their 20s and 30s put their hands up. They would have a thermode, similar to the one Wiech used in her Virgin Mary study, strapped to the inside of their left forearms. For three weeks, the volunteers trooped in and out of the hospital for the pain sessions, eight in all. At each visit, the thermode, a three-by-three-centimetre black ceramic hotplate, was switched on and off in eight-second blasts, over and over, a total of 45 times. It sounds like torture, but the discomfort is equivalent to mild sunburn. It did leave a red square on folk's arms after each session, and that shape was crucial for the next part of the study, which, to understand it fully, requires a brief return to Clifford Woolf's iconic 1983 experiment at University College London.

When Woolf pinched or scalded a small area on the rat's paw again and again, its behaviour changed. The rat began to pull its paw away, as a reflex, from things that shouldn't hurt, like the gentle stroke of a paintbrush. The technical name for this — things hurting that normally don't — is allodynia, from the Greek *allos*, meaning 'other', and *odyni*, meaning 'pain'. But something else happened, too. Incredibly, when Woolf pinched the rat's leg right up on the thigh — well away from the injured paw — it pulled the paw back, a classic pain response. The technical name for this — things hurting more than they should *away from* the zone of injury — is secondary hyperalgesia, i.e. excess pain, from the Greek *hyper*, meaning 'over', and *algia*, another word meaning 'pain' (primary hyperalgesia is excess pain when the zone of injury itself is stimulated). Both allodynia and secondary hyperalgesia come about because of those changes in the spinal cord that Woolf called central sensitisation.

You can probably see where this is going. If you burn a person over and over, they are likely to get central sensitisation. The question is, how are you going to know that for sure? Well, you can use something like secondary hyperalgesia as a stand-in. Which is precisely what Salomons' team did. At the end of the first and eighth sessions, a researcher kept each volunteer back and located the burn from the thermode on the volunteer's forearm. Next, they armed themselves with the scientific equivalent of a sharp pencil, a nifty thing called a von Frey filament — it's like a bit of stiff nylon fishing line that, when you press it into the skin until it starts to bend, delivers a uniform, mildly unpleasant prick. The researcher started out wide, poking the filament into the volunteer's wrist then bringing it closer and closer to the red square of the burn. 'Tell me when the pinprick feels more intense,' they kept asking. When the volunteer said 'now', they made a pen mark on the skin. Then they did the same thing starting at the crook of the volunteer's elbow, then the side of the volunteer's forearm, repeating the outside-in pricks over and over in a pattern resembling the stripes on the Union Jack. When they finished, the researcher had mapped out an oval around the burn. Aside from the red square at its centre, the oval contained only fresh skin, never in contact with the thermode, that had become painful to touch. This was classic secondary hyperalgesia. When Salomons subtracted the red square from the big oval, he was left with the exact area of extra pain sensitivity, in square centimetres, that the person had developed across the eight burn sessions. He now had a measure of central sensitisation in a single, unambiguous number.

You will understand by now that Salomons had to have another trick up his sleeve. Before the entire show kicked off, the volunteers, effectively at the flip of a coin, were split into two groups. One group sat down for some nice chats, centred around scenarios designed to beef up their interpersonal skills. Imagine, asked the researcher, that your best friend comes over to discuss a personal issue at a rather inconvenient time — you're about to go to bed. How might you handle that? I'll let you ponder that one. The point was, it had nothing to do with pain, it was merely a control for the real therapy the other group was getting. Those folk got a short, sharp session of cognitive behaviour therapy (CBT) to hone their pain-coping skills. This involved influencing the volunteers' beliefs about pain by offering them a variety of scientific facts about pain as well as some plain truths about the experiment.

First up, CBT had to deal with Salomons' pain machine, a futuristic white box with an ominous black cable attached to the thermode. 'When they see this thing, they are kind of freaked out by it. They're like, "What is that going to do to me? Is it going to harm me? Is it going to burn me?"' says Salomons. He explained that it wouldn't cause lasting skin damage, and then the CBT took a deep dive into how the volunteers might cope with the pain sessions.

To do this, Salomons sought traction from something meaningful to each person. One participant was on a gridiron football team. 'That's a pretty high-contact sport. There's a lot of hitting and so he routinely had to live with mild but annoying pains. In that instance, we would focus on, "Okay, do you think, once you have practised this for eight sessions, it could help you cope with those kinds of pains?"' For some people, Salomons highlighted the bald fact that they were getting paid

$130 to do the experiment. It might seem like clutching at straws, but the strategy reflects, says Salomons, what happens in real psychotherapy. 'We would say, "Focus on this aspect of your negative thinking, or try to give yourself a little mantra, a positive encouraging thought. Give me a list of those thoughts so you can see which one works best."' For example, it's common in the real world for people to see pain as a sign that the body is failing. An inner pep talk, to the effect that persistent pain doesn't usually mean tissue damage, can help. The goal was to remove the threat, to reappraise pain from 'How am I going to endure another hour in Dr Salomons' torture chamber?' to something they could deal with. So what happened?

After the first session, the group taught to deal with their needy, late-night friends had an average area of secondary hyperalgesia measuring 45 square centimetres. That meant the repeated burns had created an oval of sensitised skin around the thermode square after only a single session. Over the entire experiment, their central sensitisation got worse — after the final, eighth session, the area had grown to 48.5 square centimetres. The CBT group started with a similar figure — they had 48 square centimetres of secondary hyperalgesia after the Union Jack routine was done. But during the course of their pain sessions, against which each person was armed with mini-mantras and self-sweet-talk based on facts, their nervous system took a very different path. After the final session, the area of secondary hyperalgesia for CBT recipients had not increased — it had dropped from 48 to just 30 square centimetres, a reduction of nearly 40 per cent. CBT hadn't just shifted their mindset, it had rewired their pain-sensing system. CBT had, effectively, wound back their central sensitisation.

I can't overstate how important this is. When you think about reframing pain in a positive light to close the spinal gate — so-called top-down modulation of incoming pain signals — it seems like hard work. There is a sense that you need to 'stay positive' to control your mindset and keep pain at bay. Salomons' findings suggest that's not the case. Fact-based reappraisal can have a rapid effect on pain, probably via gate closure, but it also has hard physiological effects on areas that mediate central sensitisation, like the dorsal-horn neurons. All of which adds up to 'mind over matter' being a gift that keeps on giving for pain patients.

But how does it work? Salomons did fMRI brain scans at the start and end of the experiment in both groups. The findings weren't conclusive. He points out, however, that, when you map the descending pain pathway, you find areas of the midbrain, especially a collar of tissue called the periaqueductal grey, that send signals to the dorsal horn to inhibit pain. It is an area, says Salomons, with two very important connections: 'The periaqueductal grey also gets information from regions of the brain like the amygdala and the prefrontal cortex that are in charge of our emotions and our thoughts.' The amygdala is our fear centre and the prefrontal cortex our seat of rationality, two areas in a constant battle for supremacy in pain patients. 'That creates a very plausible mechanism by which thoughts and emotions can alter how we receive incoming nociceptive [pain] information.'

It is a potent lesson. Get a mental key to see pain in a new, less-threatening light and the pay-off can be rapid. The key can produce actual changes in the nervous system, something borne out by objective measures like secondary hyperalgesia. Take on board facts that shift your thoughts away from the catastrophic and shift your feelings away from the doom-laden and you can put a brake on central sensitisation. Salomons is now in the midst of massively upscaling this smaller study, to examine people who are about to have surgery for a range of conditions and who are therefore at risk of central sensitisation. He wants to predict which of these people will benefit from CBT as a pain prophylactic.

Some people, of course, are not the type to delve into their inner psyche and motivations. They just want to get active again. So could there be a way to combat pain without targeting beliefs? Might there be a way to improve pain without thinking too hard about it?

Chapter 4

Adherence Is Critical

the importance of regular exercise

On a still autumn morning in London in 2017, Lawrie, his wife, and a couple of Australian friends strode along the pavement, the flagstones still damp with a mosaic of night-time dew. They were on a mission. Lawrie had made a list of art galleries, and they were picking them off one by one. Today it was the Tate Britain, and, although the bus had got them close, there was always going to be a lot of walking. The capital rose from slumber as the group marched at a steady slog past cafes where the chink of china signalled the day's first coffee, and a solitary black cab wisped exhaust that, along with the wet street, branded their nostrils with the distinctive odour of London town.

They weren't too far from Baker Street, whose famous former resident, a Mr Holmes, would almost certainly have noticed something curious about the group. It stuck assiduously to the footpath on one side of the road. Always the left. As the bunch zigzagged through the streets of Westminster, past regiments of squat Georgian townhouses and the red-brick Arts-and-Crafts apartments that line John Islip Street, if one member struck out across the road, Lawrie would always bring them back. 'I had to drag everybody over to my side of the road, because I needed to walk on the left-hand side,' says Lawrie, who is in his late 60s and, by his own admission, a little on the heavy side at just over 100 kilos. 'The people we were travelling with were very kind and accommodated that.'

Why did he have to walk on the left? For 18 months, Lawrie's left knee had been giving him trouble — at first a niggle, and then, increasingly, a full-blown pain that brought his six-foot frame to a hobble. Something special about the left footpath was soothing his pain. It was all to do with a design feature called cross fall. To encourage water to flow away from buildings into gutters and stormwater drains, councils mandate that footpaths slope down to the street with a gradient of between one and two-and-a-half centimetres per metre. When Lawrie walked on the left, the cross fall shifted his posture and opened up the inner side of his left knee, lightening the load at the root of his pain. Walking on the right did the opposite: it was like tramping along a steep hillside with the slope reversed, and put pressure on that exquisitely sensitive medial side. In the end, Lawrie's obsession with street polarity worked a treat, and the friends managed to tick off the National Portrait Gallery, the Victoria and Albert Museum, and a host of others. But London was just one stop on a European trip. Lawrie was about to head to a place where attention to footpath geometry was decidedly less rigorous.

They had rented a cottage in the tiny French village of Berbiguières, a postcard-pretty enclave of stone houses that fan in a half-circle below an eponymous 12th-century chateau. The whole arrangement is perched on a hillside, bounded by an uneven road that weaves serpentine around the edge, and riddled with cobbled alleyways and steps. There is no flat. Lawrie's cross-fall strategy didn't apply here, and his knee started to get worse. But there were sights to be seen. Berbiguières is in the Dordogne in south-west France, a department that maps roughly onto the old county of Périgord. The region is famous for its black truffles but also its prehistory, in particular a string of troglodyte villages set into the limestone cliffs along the banks of the Vézère river, first inhabited during Neanderthal times. Lawrie had scheduled a trip to one of those sites, La Roque Saint-Christophe, a horizontal slit-like crevasse 80 metres above the river, a vantage point that gave its inhabitants a distinct edge on would-be marauders. Access was via steps. Lots of them.

'Climbing up all of these steps took a bloody long time. I'm very lucky to have such a supportive wife and good friends who were patient with me so that I could, ever so slowly, get there. It would have been over in half the time if I'd been able to walk properly,' says Lawrie. But getting eyes on those Stone Age digs came at a heavy cost: Lawrie's knee became excruciatingly sore. A year and a half earlier, he'd been walking ten kilometres. Now he was barely shuffling 50 metres, even when using hiking poles for support. 'It was eight-out-of-ten pain, always there, searing on the inside of the knee underneath the patella, like when you fall over onto a hard surface and you're ill prepared for it. Bang,' says Lawrie. 'It was very depressing. That was the straw that broke the camel's back.'

Lawrie had to do something about his knee. They flew back home to Melbourne, and he booked in to see an orthopaedic surgeon, whose rooms were in the leafy inner suburb of Malvern. 'I hobbled in, and the surgeon said, "I can tell you why that's hurting." And I said, "How would you possibly know, we've only just met?"' The surgeon turned to his computer and flashed up an X-ray of Lawrie's knee from 18 months earlier, which Lawrie's GP had ordered when the pain started. 'He said, "I can guarantee you it's going to be worse now."' The surgeon sent Lawrie for new X-rays, and, when the pictures were processed, you didn't have to be a radiologist to see he was in trouble.

The bottom end of the thigh bone usually floats like an evenly balanced seesaw above the top end of the shin bone, the two bones separated by a thick buffer of cartilage and lubricating joint fluid. Lawrie's X-ray looked like a big kid was sitting on one end of the seesaw, and it had hit the ground. The inner side of the joint was 'bone on bone'. Lawrie had osteoarthritis, a grinding wear and tear of cartilage and bone that, fuelled by low-grade inflammation, causes pain, stiffness, and difficulty bending the knee. Worse, it was category four, the most severe. Surgeon and patient sat down for an earnest chat about Lawrie's options. He would, it was decided, need a total knee replacement, where the ends of the thigh and shin bones and the back of the kneecap are shaved off and replaced with shiny metal and plastic implants.

But there was a complication. Lawrie's retired now, but, in his heyday, he was one of the world's top viola players, sitting in the front desk of Australia's leading symphony orchestras. You don't tend to think of classical musos as overly fond of a drink, but, says Lawrie, he imbibed liberally during his concert days. Which may have been a factor in his weight gain. Pictures from back then show a full-faced, bearded man, smiling though brass-rimmed glasses and resplendent in white tie and tails, clutching a magnificent viola made in 1767 by the Italian luthier Carlo Landolfi. These days, the beard is white and the hair sparse, all scored to the tempo of Lawrie's easy smile and gentle laugh. But his face is rounder, too, and, over the years, the weight has ushered in its habitual companions, high blood pressure and diabetes, which, with high cholesterol, made Lawrie vulnerable to one health problem in particular.

'I had a routine check with my diabetic specialist, and he said, "What on earth have you been doing? Your triglyceride levels are shocking," and then he sort of blanched when he saw I was due to have a knee replacement.' Lawrie, unfortunately, ticked nearly every box for heart disease. Having a knee operation with an undiagnosed heart condition could be a risky proposition indeed, so the diabetes doctor wanted Lawrie to get checked out, starting with a test called a stress echocardiogram. You take a brisk walk on a treadmill that is gradually tilted uphill, while your pounding heart is scanned with an ultrasound. But Lawrie's best pace was a dodder — he couldn't do it. In the end, they had to make his heart beat harder with a drug instead, and, when the results came in, his ticker had made a poor showing. Segments of Lawrie's heart chambers hadn't pumped well under the strain, which meant the coronary arteries that supply the heart muscle were narrowed. Things were escalated again, and Lawrie was shunted off to a heart specialist. The cardiologist didn't mince words. What Lawrie needed, he said, was an angiogram. They would inject dye into his coronary arteries, and, if blockages were found, the arteries could be opened up with a stent.

Lawrie remembers lying in the hospital bed after the angiogram, his wife sitting beside him, when the heart doctor came in to tell them how it went. 'My wife said, "Did you do the stent?" He said, "Actually, no, we didn't. Your husband is very sick. He needs a quadruple bypass. Now." To put it in blunt terms, I was a bee's dick away from dropping dead. He said, "Unless you've got a real reason to go home, you're staying in hospital."'

It was time to prioritise, and, in the hierarchy of survival, heart trumps knee any day of the week. So plans for Lawrie's knee replacement were shelved, and, in March 2018, he underwent surgery to transplant strips of his forearm arteries across his corroded coronaries to restore blood flow to the heart. The operation went without a hitch, but Lawrie was about to confront another hurdle that would push his resolve to the limit. Cardiac rehabilitation is all about getting the body fit after surgery, and it centres on a strict program of exercise. But Lawrie had a bung knee. He couldn't even walk on a treadmill. How could he possibly do it?

'I moved forward to cardiac rehab, and that's when I ran across Milly. Or rather, she just ran right over me and said,

"Okay. Start exercising. This is the way it goes." She's not brutal, but she's very demanding, and she's tough and doesn't take any backwards steps. She keeps pushing.' Lawrie saw Milly, a physiotherapist who was doing a PhD on osteoarthritis and exercise, twice a week for one-hour sessions. They started gentle with stretching, walking, and Lawrie rolling his legs over on a stationary bicycle. Even so, his knee hurt, and he was limping. Then Milly asked him to step up and down on a low platform. Lawrie did it, just — each rise was more of a totter than a step. So Milly had him sit down on a chair and lift his sore leg until it was straight. Then she wrapped a kilo weight around his ankle and had him lift the leg again, bumping the resistance up steadily over each session until he was lifting five kilos. Lawrie did ten repetitions of each exercise, in sets of three.

When the six-week program was up, something quite unexpected had happened. 'I said to her, "Milly, I gotta tell you, I hate exercise, I hate gyms, but I'm going to sorely miss this class,"' Lawrie recalls. 'Because at the end of it, after doing all of the load-bearing stuff and knee lifts, all of those exercises, I started to be able to walk without a stick. I said, "I really notice there's a huge difference going on here. I'm able to walk — not fast — but I'm walking."'

It was real progress. But Lawrie's knee still hurt. What he needed, Milly decided, was more exercise, of a specialised kind. So she referred him to the optimistically named GLA:D, which stands for 'Good Life with osteoArthritis: Denmark', a program of education and exercise that began in Denmark in 2013 and is tailored to people with hip and knee osteoarthritis.

In the chill Melbourne winter of 2018, Lawrie turned up for his first session at the rehab hospital. Now, the GLA:D moniker might sound overoptimistic, but its reality under Lawrie's new taskmaster, a physiotherapist named Jason, was simply pragmatic. Lawrie's repertoire of exercises grew. Again, he started slow, standing from a chair and doing a series of step-ups. Jason urged him on. Lawrie executed forward lunges with the weight on his sore knee, sideways lunges, single-leg squats, sit-ups, and pelvic lifts on an exercise ball.

For the next step, Jason introduced TheraBands, oversized elastic bands that are colour-coded, from super stretchy tan, which offers a kilo of resistance, to gold, which serves up around six kilos at full stretch. Lawrie sat in a chair with the band looped around the chair leg and his ankle, then lifted his leg until it was straight, drawing the band out to its full length. He did it over and over. Then Jason anchored the TheraBand to the wall, and Lawrie moved his leg forwards, backwards, sideways. Ten times in three sets. Always three sets. His efforts began to bear fruit. 'I'd hobble in initially with maybe six-out-of-ten pain, and I'd walk out feeling, initially, maybe only five. But after about two or three weeks, I walked out with almost no pain. It just felt stronger. I felt stronger, period.'

For six weeks, Lawrie thrust his exercise phobia to one side and attacked the GLA:D schedule with zeal. The whole thing took him back to the days of drilling through études and scales on his viola and, when he'd joined the Australian Broadcasting Commission training orchestra, the endless repetition of rehearsal, mostly in the cavernous interior of the Lindfield

Masonic Hall, a red-brick Art Deco building in northern Sydney. 'You sit down, you practise and you practise and you practise, and, when you think you've done enough practice, you do it again. You keep doing things till you've learned them or they're completely in your control. I felt a bit like that with the exercises. You have to do the exercise, you have to keep doing the exercise, and then, when you think you've done enough, you have to do it again.'

In July 2018, Lawrie graduated from GLA:D, and Jason ran some numbers on him. They make for remarkable reading. Lawrie's knee pain before starting GLA:D averaged eight and a half out of ten, and he was popping slow-release paracetamol (acetaminophen) three times a day. When he finished GLA:D, his pain had tumbled to just two out of ten, and he wasn't taking any pain meds. Jason brought him back a year later to see how he was faring. Lawrie's average pain was three out of ten, and he only needed paracetamol around three times a week.

In September 2019, nearly two years since Lawrie's agonising sojourn to those Périgord caves, he decided it was time for another trip. At the toe of Italy's boot, like a crumpled football about to be kicked, lies the island of Sicily and, on its southern coastline, abutting the emerald waters of the Mediterranean, the ancient town of Agrigento. The town drapes across a plateau bounded by low cliffs and, like much of Sicily, is built on many levels, which are, of necessity, linked with steps. 'My god it is the island of steps. Everywhere, there are steps,' says Lawrie. 'The more I did steps, the stronger my legs became.'

Agrigento's classical roots go back a long way. It was settled by Greeks during the diaspora that preceded the Hellenic Golden Age, around 500 BC. A legacy of that time is a clutch of temples in the Doric style, dedicated to Heracles, Zeus, and the god of medicine Asclepius. These are scattered across an exposed ridge just to the south of the town, known as the Valley of the Temples. It's an international tourist drawcard, and it is vast, covering more than a thousand hectares. On size alone, that's enough to strike fear in the heart of anyone with a dicky knee. Lawrie, however, was unfazed. 'It's a whole day's walk. I can't begin to tell you. It's one complete, ancient city of ruins and temples, and you just walk and you walk and you walk all day.' How did he handle all that walking? 'Yeah. Not a problem.'

Lawrie had also, of course, been on an unplanned trip with a less-welcome itinerary, one that included pain, trepidation, open-heart surgery, and a knee replacement. Except he never did make that last stop. Incredibly, his dogmatic adherence to an exercise regime looked to have somehow fixed his knee pain. But Lawrie had seen the X-rays with his own eyes. His knobbly old knee was, as they say in the vernacular, cactus. How could exercising on a knee deemed so decrepit that it was to be consigned to the biological trash can make it better? What was going on?

There's an unsightly grey circuit board in a room of Kathleen Sluka's house in Iowa City, a college town set amid rolling croplands of corn and soybean in the Midwestern US state of Iowa. Not one to sit idle, Sluka set to correcting the aesthetic breach. Armed with brushes, acrylic paint, and a canvas, she began applying a select range of hues from her palette, ultramarine and Prussian blue, vermilion and orange and cadmium yellow, mixing as she went. The resulting picture is oddly intoxicating. Bulbous tubers in red and blue push out stems, rhizomes that wind and fork to make an organic frame, through which another, very different structure is lit. Regimented lines expand abruptly into nodes, their serried ranks rotating left and right in an unending repetition of some sterile order. It is Sluka's very own portrait of the meeting of human brain cells and the man-made attempt at their equivalent, the silicon chip. And it is all now mounted on vertical hinges that swing shut over that ugly circuit board in her studio.

You can see Sluka's prolific and stunning array of art on her website. It's a riot of colour, all riffing to the same themes: the brain, nerves, and muscle cells. Why the obsession with painting biology? It's not just about prettying up a wall. Sluka is one of the world's leading pain researchers, and, three decades ago, she faced a conundrum that meant that taking a close look at those human cells simply had to become her life work. In the process, she stumbled across something that would turn our understanding of the role of exercise in pain on its head.

In the mid-1980s, Sluka was fresh out of physiotherapy school and working in a clinic in Houston, Texas. A large number of her clientele had back pain, and Sluka had assembled an armoury of tools to deal with them. She would lay on hot packs, tingle their skin with a TENS machine, pulse their back with an ultrasound probe, and have them kneel facedown on a mat, arms splayed forward like supplicants, to stretch their back muscles. By and large, it worked, and patients would come back for more, as often as three times a week. Now, depending on your professional goals and business model, that might sound like success. Not for Sluka. 'I thought, "This isn't right, we need to get them to take care of themselves,"' she remembers.

One patient in particular sticks in her mind. She was a middle-aged woman, a little overweight, with back pain from no apparent cause, for whom Sluka had prescribed an exercise program. Nothing too challenging. Just regular walking, sit-ups, a few arm and leg raises on the mat while on all fours. 'Very simple stuff,' says Sluka. 'But just enough to get her moving.' Simple maybe, but it worked a charm. 'Her pain would actually go away completely when she exercised.' But the woman kept coming back every six months, like clockwork, with more pain. 'I'd say, "Are you still doing your exercises? And she'd say, "No. You mean I have to do them for the rest of my life?"'

Did she? Well that all depended on how exercise was helping her pain, a riddle to which Sluka had no answer. She had to find out, and that would mean leaving the physiotherapy clinic behind. As it happened, the world's top pain-research institution was just a few kilometres up the road, across the causeway on Galveston Island at the University of Texas. So Sluka signed up for a doctorate in anatomy and neuroscience, took to the lab with gusto, published a swag of papers, and, in short order, landed a professorship at the University of Iowa, another top-tier research institution. Its historic campus straddles the Iowa River and centres on the Pentacrest, a cluster of buildings that surround the Old Capitol, a Greek-revival pile that was once the seat of Iowa's government. It was all to Sluka's liking, and she found her feet quickly, pushing on with research to tease out the architecture of pain in dozens of experiments. But the enigma of exercise sat stubbornly on the backburner. Until one day, a hopeful PhD student wandered onto campus. Her name was Marie Hoeger Bement, a physiotherapist working with chronic-pain patients. And she wanted to study exercise.

The pair hit it off and, with their shared experience of the pain clinic, had both puzzled over the same quandary. They knew that, when we exercise, our brains squeeze out endorphins, the body's trademark painkillers, which work on the same receptors as drugs like morphine. But something about endorphins didn't fit with what Sluka and Bement had seen with their patients. The research showed that you had to exercise hard before endorphins got pumping — an effort equivalent to 45 minutes of a dance workout, a full hour of running at a marathon pace, or cycling at around 85 per cent of your maximum power. But the two had seen patients benefit from walking and mat exercises at nowhere near that intensity. They wanted to know if low-level exercise really did help pain, and, if it did, could endorphins be the key?

They decided to study rats, whose uniform genetics, similarity to human physiology, and ability to cope with the controlled environment of the lab gave the best chance of producing a meaningful answer. Their first challenge was to have the rat mirror the human trajectory of chronic pain, which, for almost every person, begins with the blissful state of no pain at all. Sluka and Bement needed to know the animals' pain threshold in this normal state. So they employed a sheaf of von Frey filaments, which, you might remember from Tim Salomons' Union Jack study, are strands of nylon that you press into the skin. They pushed the nylon into the hind paw of each rat, starting with skinny, super-bendy, not-very-painful filaments and progressing up to thicker ones that could deliver a pain punch. The filament with enough force to make a rat pull its paw back was recorded as the animal's everyday pain threshold.

The next step was to, humanely, have the critters enter the wretched world of pain. The researchers put each animal under anaesthetic and injected acidic salt water into the rat's calf muscle, repeating the injection five days later. The shots cause enduring pain without inflammation, so they're a good stand-in for what's experienced by people with chronic pain and no tissue damage. Sluka and Bement ran each rat through the bendy-filament test again, and, sure enough, the creatures pulled their paws back to a much less painful prick. The rats officially had mechanical hyperalgesia — more pain from a sharp jab.

Now it was time to get moving. Rats can reportedly hit a top speed of 13 kilometres an hour. The pair had them hop on a treadmill turning at a layabout's pace of a third of a kilometre per hour, for 15 to 30 minutes a day, over five days. The results were immediate, and breathtaking. Within minutes of leaving the treadmill on day one, the rats' pain thresholds had returned to normal, and, when the two researchers dutifully applied the von Frey filaments over the next five days, the overall effect held. It was a vindication of everything they had seen in the clinic. Low-intensity exercise had unwound the rats' pain. But how?

To find out, Sluka and Bement had paid special attention to one group of rats. Before each day's stroll, they injected naloxone into these rats' furry abdomens. It's a drug I administered hundreds of times over the years in the emergency department, mostly to people who had overdosed on heroin. The effects are dramatic. Someone who is blue and not breathing is awake in a matter of seconds. Naloxone stops heroin attaching to opioid receptors, and it does exactly the same thing with our endorphins, blocking the feel-good effect. How did the naloxone rats do after their treadmill sessions? Horribly. With their endorphins neutralised, exercise did precisely squat as a painkiller. The pair had made an important finding in the annals of pain research. Not only did low-intensity exercise help with pain, it worked, at least in part, by switching on endorphins.

Bement got her PhD and moved on to a new lab, while Sluka pressed on with her research, steadily amassing an impressive list of publications. Then, in 2011, something strange happened. Sluka had been trying to get data on a routine question: how far will mice run after a calf injury? First, she needed to get the mice used to running on a wheel, so she asked her research assistant, a woman called Lynn Rasmussen, to give them five days of free rein, with the running wheel hooked up to a counter that totted their mileage. At the end of the five days, Rasmussen injected acidic salt water into their calf muscles to give them hyperalgesia — the proxy for chronic pain. But something was off. 'I asked Lynn to follow up and see what happened. But the animals didn't develop their hyperalgesia. They didn't develop anything,' says Sluka. 'I'm like, "Oh, what's going on?

Okay, that's weird. Do it again."' Rasmussen did the experiment again. And again. Each time, the result was exactly the same. The mice weren't getting chronic pain from the injections. Sluka had induced the pain model in hundreds of rodents without fail. Why wasn't it working now?

One of Sluka's paintings features a bunch of pompoms, purple, blue, green, and gold, big ones and small, an amorphous cluster that two slender black hands are trying to gather up, like someone making beanies had lost their load while coming back from the haberdashery store. Sluka, of course, paints biology and these are not craft supplies but cells of the immune system. One of those cells was in her sights in the mystery of the pain-resistant mice. It's called a macrophage. 'There are resident macrophages that sit in your muscle. They sit there, they police the micro-environment, that's a normal thing that happens,' says Sluka. If your muscle gets injured, it's macrophages that surge in to direct repair operations. In the process, wondered Sluka, might they be a factor in the pain equation?

Macrophages, she knew, have a Siren-like relationship with a bumpy, orb-shaped cell called a liposome. Macrophages simply can't help themselves around liposomes — they have to ingest them, but, in a case of unwitting suicide, the macrophages promptly die. Sluka injected liposomes into the rats' calves to lay waste to all the macrophages, and discovered that, when the rats got their acid injection, they didn't get any pain. Macrophages must have been doing something to trigger the pain response. But what?

Sluka knew that answering that question could solve a

problem that plagued every pain therapist. Because while many chronic-pain patients improve with exercise, others don't. And some actually get worse. What Sluka needed was an animal model that mimicked the experience of this latter group — particularly the unfortunate chronic-pain patients whose pain got worse after a single session of exercise. So she tweaked the injection, making it slightly less acidic: on its own, it didn't hurt. Then she added the equivalent of a seriously fatiguing exercise. She had some animals run on the wheel for an hour or so, and she would electrically stimulate the calves of others — six minutes on, six minutes off. Just like her former patients, the animals developed pain from a single workout. Sluka now began a series of experiments to understand why, all driven by some bedrock understanding of the biology of exercise.

Let's say you go for a run. Your leg muscles start working hard, sucking in oxygen, which mixes with glucose to make something called ATP, the equivalent of an energy bar for your body's cells. Keep running and your muscles go anaerobic, burning through glucose in the absence of oxygen, whose supply has run dry. The result is a build-up of lactate — the runner's nemesis that makes legs burn — and a more acidic environment in the muscle.

Where do macrophages fit in? Macrophages, it turns out, are exquisitely sensitive to the things that muscles produce when you exercise hard: ATP and lactate. In an extraordinary result, Sluka's research found that macrophages respond to these products by making pro-inflammatory cytokines, which rankle the muscle's nociceptors. Fatiguing exercise was prompting macrophages to spit out pain-pinging proteins.

But there is something crucial to remember about the mice in these experiments. They were modelling pain from a *single* session of exercise. How much exercise had they done before that solitary outing? Zilch. These were typical lab rats, idlers spending their days doing nothing, or loitering passively when the urge took them. Not unlike the couch potatoes in our own population, when you think about it. What would happen, Sluka pondered, if you gave those animals the equivalent of an unlimited gym membership? What would happen if her mice got ripped?

So Sluka put running wheels in the animals' cages. 'They could go in there and run at will, whenever they felt like it, and mostly they run at night,' says Sluka. 'Anybody that has had a gerbil has seen this before.' Some rodents ran over five kilometres in a night, though not all took to the wheel with gerbil-like alacrity, some running as little as 200 metres. 'They're kinda like people. Some people are super active, and some people aren't so active.' There was, however, a constant. The animals were exercising, not just as a one-off, but regularly, for anywhere between five days and eight weeks. And when Sluka gave these animals acid injections, something very special happened. Or rather, didn't happen. They didn't get any pain. 'As long as they ran, it didn't seem to matter whether they ran one kilometre, five, or more. There was no correlation with the amount of pain relief based on the amount of running-wheel activity.'

We're talking on Zoom, and I'm looking intently at Sluka. She's fair-skinned with grey-green eyes and strawberry-blonde hair that feathers around her shoulders, kept in check by the band of her headphones. She's getting voluble and a little flushed now, her skin standing out pink against her preferred Zoom background, the blue-green of Monet's water lilies. 'We had been doing experiments in sedentary animals all this time,' she says. The acid injections had always caused pain because none of her animals ever moved. Get them fit and they became, in effect, resistant to chronic pain. Exercise, she had found, didn't just treat pain, it could prevent it in the first place. Exercise was the good guy after all, but you needed to earn loyalty points. You needed to keep coming back.

Something, though, still didn't make sense. Where were the macrophages? Why weren't they ringing the pain bells with all that exercise? Sluka took another look, taking a sample of the animals' calf muscles and staining them to show up the macrophages. Something profound had happened.

Macrophages, it seems, come with something of a split personality. When you're injured, they're whipped into a frenzy, pumping out pro-inflammatory cytokines to mobilise the battalions of cells needed to get healing underway. These macrophages have a special designation: M1. But for healing to proceed smoothly, inflammation has to be turned off. It happens that the Yin of the M1 macrophage has its very own Yang, in the form of an M2 macrophage, which does just that: M2 macrophages halt the incoming immune cells with a barrage of anti-inflammatory cytokines.

When Sluka looked at the calf muscles of rats that had been sitting around, there was a preponderance of M1 macrophages.

But when the rats got fit, their muscles were packed with M2 macrophages. Regular exercise had prompted individual macrophages to undergo a personality switch, from Hyde to Jekyll, with far-reaching implications for pain. Exercise still produced ATP and lactate, and those things still triggered macrophage activity. But the macrophages were different beasts altogether.

For exercise to be anti-inflammatory, the dose is critical. If there is a lesson for people with persistent pain, it's that the first session of exercise, especially if you're in poor physical shape, is likely to hurt. 'Yet if they do it routinely over time, there's at least one study that shows that pain during activity actually goes down. So after two or three weeks, if you can stick with it, you'll have less pain during activity, but, more importantly, you'll have less overall pain,' says Sluka.

If you're like me, though, you'll be troubled by a big question. Exercise prevented pain from an injury, but one of the main reasons that pain exists is to stop us moving so injuries can heal. Why would another system evolve to suppress pain and keep us moving on the injured part? 'Movement helps promote the proper laying down of collagen,' says Sluka. Collagen is one of our main structural proteins, and it changes orientation, or remodels, during healing. Movement can ensure the final configuration is the one that functions best. But there is another reason. 'The normal state is being physically active. The abnormal state is being sedentary. So evolutionarily, we already had a state that kept minor injuries at bay so that they weren't pain-producing. You could continue to do things if you had a small injury.' This theory, Sluka concedes, is pure speculation, but it does speak to our intuition that movement was key to survival in the swamp of prehistory. 'Cavemen were out killing things and getting food. They weren't sitting around at desks talking to people on Zoom all day.'

Since the key to reducing pain through exercise is regularity, Sluka advises doing something that you like. 'The best thing for someone to do is to do what they will commit to. If you like yoga, do yoga. If you want to go for a walk in the woods, go for a walk in the woods. If you prefer walking on your treadmill, that's fine. If you want to go swimming, great. If you hate swimming, don't do it, do something else. The number-one reason why exercise doesn't work is because people don't do it. Adherence is critical.'

Sluka's work in animals is rigorous. The sheer number of scientific studies she's carried out to establish cause and effect in exercise and pain is testament to that. But when it comes to prescribing exercise to people with chronic pain, the whole field is saddled with a depressing bugbear.

Tour around the low-lying island of Funen in Denmark and you'll be greeted with a surfeit of spire-encrusted gothic castles that look straight out of a Hans Christian Andersen fairytale, which they may well be, because the writer was born here in 1805. In contrast, the main building at Odense University Hospital is a dramatic, modernist, canoe-shaped obelisk of towering blue glass, dwarfing the traditional, rectory-style red-brick houses that line the neighbouring street. It seems unsettlingly dystopian, not unlike Andersen's darker stories, and it houses its own tales of woe. The hospital is home to the Pain Centre South, a clinic that sees a staggering 1,300 new chronic-pain patients each year.

Back in 2010, the clinic welcomed a young physio who wanted to play his part in easing that woe. He had the lean physique and ragged beard of a polar explorer, and kept in shape with a punishing schedule of hikes and road-bike trips that would take him from the spruce woods of Germany's Harz mountains, to the rock and ice of the Norwegian fjords. Henrik Vægter was also a zealous spruiker of exercise for his pain patients, an approach that dovetailed nicely with a program the Danish government had set up a couple of years earlier. Coordinated Efforts for Work Retention had a very specific brief: to get people with muscle and joint pain back to work more quickly after taking sick leave. Here's how it worked: they would get state-subsidised access to a doctor, psychologist, and physiotherapist, but, in return, they had to sign up to a gym and do regular exercise. The early results of the program were encouraging. Work absenteeism dropped 34 per cent, with participants reporting less pain and better function. Vægter, however, noticed a disturbing trend among his patients in the program.

'I had so many patients tell me, "This makes my pain worse. I cannot exercise. When I exercise, it just flares up my pain,"' says Vægter. He scoured the literature, and his puzzlement grew. An impressive body of research told him his patients should be getting better with exercise, not worse. The HUNT study, for example, published in 2011, questioned more than 46,000 Norwegians and found that 29 per cent had chronic pain. But in people who exercised for at least 30 minutes three times a week, the prevalence of chronic pain fell by 12 per cent across the board, and by nearly 40 per cent in older women. Why weren't his patients getting better with regular exercise?

Vægter started digging and came across a phenomenon called exercise-induced hypoalgesia, or 'less pain with exercise'. As in Sluka and Bement's mice that got pain relief on a treadmill, exercise-induced hypoalgesia happened fast, sometimes in minutes, but this time in people. Yet many of Vægter's patients were defying the trend. Vægter wanted to know why, and he began to study the phenomenon intensely.

His experiments started out with healthy subjects. Vægter would have them relax in a chair, and then he'd bring out a device called a pressure algometer, which looks like a 3D-printed handgun, with a stubby rectangular grip and a tubular metallic barrel. Press the device on a surface and it registers the exact force on an LED readout. With due warning, Vægter would push the algometer into each person's thigh. 'Tell me when the sensation changes from pressure to pain,' Vægter asked, jotting down how much force was needed to cause pain. This was their pain threshold. Sometimes he'd keep pressing, to find the maximum pain the person could take, recording this as their pain tolerance. Figures in hand, Vægter ushered each participant to a nearby exercise bike for a 15-minute ride, just enough to put colour in their cheeks and have their legs feeling it, and then he took the algometer readings again. Vægter ran person after person through this routine, and the effects were remarkably consistent. 'We would quite robustly see an increase in threshold and tolerance after bicycling, meaning they could endure more pressure before it became painful, and more pain overall.' How much more pressure and pain? A full 15–20 per cent.

Exercise-induced hypoalgesia, Vægter had confirmed, was very much a thing — in healthy people. But then he ran some of his chronic-pain patients through the same experimental set-up. When those folk stepped off the bicycle, the algometer was picking up something quite different.

'There are individuals with chronic pain who have the same effect as healthy participants. But we also see that people with widespread pain in several areas of the body — which could be fibromyalgia — have the opposite response. Either they don't have the hypoalgesic response, or they become more sensitive after exercise,' says Vægter. Not only were those people denied the painkilling effects of exercise, but some were actually getting worse. How could this be? Vægter decided to home in on something that might just be hidden in plain sight.

He assembled a team of pain researchers and rehab experts in Odense, and recruited 96 adults with lower-back pain. Each had their pain threshold measured with the algometer, first pressed into the flesh of their lower back — the source of their pain — and then pressed into their calf muscle — which they would be exercising. Then they had to do a tedious drill at the local rehab centre. Their task was to walk as quickly as they could between two cones, set 20 metres apart, and cover as much distance as possible in six minutes. 'Keep up the good work! You are doing well!' yelled the researchers — well-rehearsed cheers to egg on each person as the minutes ticked by. Despite the encouragement, some participants weren't doing well at all; as they shuttled between the cones, these participants, 27 to be precise, got a nasty flare-up of their back pain. At the end of it all, Vægter's team rechecked everybody's pain threshold, pressing the algometer into backs and calves once more. People whose backs were fine throughout the ordeal got the expected painkiller effect from the brisk walk. But when Vægter analysed the folk whose back pain had flared, he found the opposite: their pain threshold went down.

At first take, that might seem blindingly obvious; hurting your back causes more pain in your back. Until you realise something else. Pain thresholds didn't just drop in their backs, but way down in their calves, too. Doing an exercise that hurt one part of the body, Vægter had found, could undo the painkilling effects of exercise in another part of the body. But there was a flipside. For those whose backs coped well with the shuttle test — 70 per cent of the volunteers — the painkilling effect occurred not only in their calves, where all the work was being done, but also up in their backs. Exercising parts that *don't* hurt creates an all-over effect. The results offer critical insight when tailoring exercise for someone with pain. Walking helps a lot of backs, but, if walking hurts, a therapist can devise a plan to exercise other muscles — perhaps through cycling or an upper-limb workout — which may still help the back.

But Vægter noticed something odd as he studied more and more people. 'The exercise-induced hypoalgesic response that a person has on a Monday morning is not very reliable. The response you will have if you come in on Monday the following week may be completely different, even though your pain situation is the same,' he says. It was a finding that didn't make sense. The pressure algometer has what's called high test–retest reliability. Press the flesh with it again and again on the same person, under the same conditions, and you generally get the same result. The Monday-morning effect was not caused by the handgun misfiring. Yet even when pain levels stayed the same, or a person had no pain at all, the analgesic effect of exercise varied from week to week. What was happening on Mondays? The experimental set-up was fixed, the subjects carefully selected. Could it be something to do with the researchers themselves?

Vægter put together another research team, and they posted flyers on a couple of university campuses in Odense. He wanted to enlist people with no pain who could hold a wall squat for three minutes. Now, if you've ever done one, you'll know that is quite a big ask. You have to bend your knees to perpendicular with your back flat against a wall, and stay there. Pretty soon, your quads are burning with the effort. That said, it is a reliable way to bring on exercised-induced hypoalgesia.

Eight-three folk made it through selection, and Vægter's team promptly set about measuring their subjects' pain thresholds. They pressed the flesh in two spots: the quadriceps, at the front of the thigh, and the trapezius muscle, which runs down the side of your neck. Then they split the squatters into three groups and gave them some bespoke information. The first group was told that exercise is a pain reliever that helps you do better on the algometer test. The second group was told that exercise can hurt and can make you do worse on the algometer test. The third group was told nothing about exercise or pain. Then the subjects were all sent to wall-squat hell and, when it was done, retested.

The study, published in 2020, was called 'Power of Words'. Here's why. People who got the positive message about exercise could take 22 per cent more pressure before they felt pain. They got a great analgesic effect from the wall squat. People who got no information also got analgesia, though not quite as much as the positive group. And the negative group? Not only did they fail to get any painkilling effect from the wall squat, their pain threshold actually dropped by 4 per cent — they got worse. What Vægter had found was that negative words were hurting people — literally. But he'd also discovered something else. Positive messages were not significantly better than telling people nothing at all.

What has this got to do with the Monday-morning effect? When someone feels pain after a workout in one week, it colours their expectations for the next week. They believe it will hurt, and so it does. But there's no point handing them a pair of rose-coloured glasses. 'What most people don't like is if we say things they don't experience. So if we say, "Exercise is a painkiller, exercise will make you happy, exercise is gonna make you fit," if that is not what they experience, it is counterproductive,' says Vægter. Far better, he says, to simply avoid the negative. Messages as simple as 'every step matters' and 'it's never too late to get started' could well be the best medicine.

The work of Kathleen Sluka and Henrik Vægter has a lot to

say about Lawrie's story. All that exercise probably got Lawrie's endorphins going. It may have produced an anti-inflammatory effect by bumping up his M2-macrophage count. And it could have set off exercise-induced hypoalgesia, pushed along by his glass-half-full attitude. Nonetheless, something about Lawrie's recovery is bewildering. Osteoarthritis is your classic degenerative disease — medical code for 'it is going to get worse'. That's just an observable fact. Wear and tear, exacerbated by one of Lawrie's cardinal complaints — being overweight — meant the outlook for his knee was inexorable, and downhill. His X-ray, remember, was 'bone on bone'. Exercise seemed, however, to have averted that apparent destiny. But how could exercise fix pain in a knee that is collapsing before your eyes?

Trace the slender thread of Melbourne's Darebin Creek northwards, and its close quarters gradually expand to open parkland and, further on, a large billabong with stands of native rushes and banks that tumble with kikuyu grass. From there, it's but a short doddle to La Trobe University and the low-rise, cream-brick-and-concrete building that houses the office of physiotherapist and researcher Christian Barton. Barton is tall, in his mid-30s, wears denim jeans and a sky-blue shirt, and has his hair gelled to natty points, evoking a kind of buttoned-down Johnny Rotten. He chooses his words carefully, in authoritative, considered phrases, but also, in the vein of that spitting man with green teeth, harbours a healthy disrespect for the status quo.

A few years back, Barton was playing Aussie Rules football

and broke one of his forearm bones — the ulna — which a surgeon repaired with a metal insert. Post-op, the surgeon advised Barton to lift nothing heavier than two kilograms for 12 weeks. Barton reviewed the literature and concluded that, for the bone to remodel, it needed loading. So he devised his own rehab plan, which included a schedule of gradually increasing weights. Within weeks, he was lifting 15-kilo dumbbells. Seven weeks after the injury, he played footy in his local league's grand final — and lost, although the result is neither here nor there. Barton's message is not to ignore medical advice, but he is hoisting a red flag on some stubborn shibboleths that dictate treatment, especially in osteoarthritis.

In 2017, Barton was one of three project leads that kicked off the GLA:D program in Australia. The GLA:D remit is very simple: to get education and exercise to people with knee and hip osteoarthritis. Barton and a small army of tutors are training up physios across the country to deliver GLA:D and then collect data on how their patients do. When we speak, Barton has recently put the finishing touches on the 2020 annual report. GLA:D, it seems, is making waves. Over three years, 7,641 people have taken the course, with 80 per cent sharing Lawrie's complaint, osteoarthritis of the knee. A year after the course, those folk reported an average pain reduction of 31 per cent. Half said they used 'less or much less medication', such as paracetamol, anti-inflammatories, or opioids. Walking speed went up 14 per cent, and joint-related quality of life — which measures the impact of knee issues on a person's lifestyle — improved 36 per cent. But GLA:D, in a dramatic reflection of Lawrie's story, also turned up something else. Before starting the program, a quarter of people with knee osteoarthritis wanted surgery. A year after finishing GLA:D, two-thirds of them hadn't had the operation. They simply didn't want it anymore.

'Most patients, and health professionals for that matter, believe the key thing in osteoarthritis that drives the pain is the structure of the joint. They get very focused on the X-ray,' says Barton. 'What we do know is that the structure of the joint relates really poorly to severity of pain, or even the presence of pain.' Getting down to 'bone on bone', says Barton, makes it more likely we'll get pain, but doesn't guarantee it. Like a barrister prosecuting a case, Barton waves an exhibit. There is an 82-year-old man at his clinic with severe osteoarthritis of the knee. On X-rays, his knee joint is narrowed to a crack. Somehow, this gent runs 15 kilometres every fortnight and does the five-kilometre community parkrun on Saturdays. 'He has no pain in his knee,' says Barton.

To understand how that's possible, it's worth hearing testimony from a man named Scott Dye. A portly man who speaks in a hurry, Dye is an orthopaedic surgeon and associate professor at the University of California, San Francisco. He's also famous for doing something extraordinary. In the late '90s, Dye cemented his place in the pantheon of doctors who put themselves at the sharp end of their experiments and give the collective knowledge of humanity a leg-up in the process. Dye, who was 46 at the time, asked his then colleague Geoffrey Vaupel to do an arthroscopy on his knee. The procedure was to have two points of difference with the everyday variety. Incredibly, Dye would forgo the anaesthetic and remain fully alert throughout. The other departure from standard protocol was a small spring-loaded probe attached to the end of the metal arthroscope that goes into the knee. The probe was specially designed to press, with a force of half a kilo, on whatever surface it touched.

Things got off to a poor start. Vaupel had to make two incisions at the front of the knee and then open a tract into the joint, pushing the arthroscope through the synovium, the pink, glistening membrane that lines the joint. That initial thrust, Vaupel and Dye reported in the paper they later wrote, led to 'involuntary verbal exclamations' from Dye, which nearly ended the experiment. The nature of those utterances remains a mystery, but Dye had confirmed that the synovium, replete with pain nerves, is exquisitely sensitive.

Then Vaupel pushed the probe deeper, and the ambience in the operating theatre dropped to a hush. Vaupel pushed under the kneecap. Dye couldn't feel anything at all. Vaupel pressed on the inner edge of the meniscus. No pain. Then he pressed on the outer edge of the meniscus, which has a nerve and blood supply — mild to moderate pain. Finally, Vaupel pressed on the lower ends of the femur and the upper part of the tibia, the parts ground down to 'bone on bone' in Lawrie's X-ray. Dye's response? No pain, just slight discomfort on the top end of the tibia. Why? The silky-smooth cartilage that lines the knee joint has no nerve supply. Grinding through it, in and of itself, cannot hurt.

So if you have knee osteoarthritis, where does the pain come from? There are, says Barton, several suspects. Bruising of the underlying bone, which can show up on an MRI, is linked to pain. Then there's the super-sensitive synovium, which can get inflamed and make pain spike. The synovium also secretes fluid that can build up behind the knee in what's called a Baker's cyst, which hurts when you bend and straighten the leg. Pain in osteoarthritis is often quite generalised, so it might come from the muscles around the knee. It could also come from the junction where ligaments attach to the bone. But Barton thinks one culprit deserves special mention.

'There's not a single nerve fibre that says, "Hey, there's pain there." There are nerve fibres that tell us information about temperature, about pressure, about movement,' he says. The thing is, osteoarthritis of the knee comes with low-grade inflammation that generates heat and fluid, so those nerves get busy. And when there is a lot of information going down the wire, mistakes happen.

A couple of years back, Barton had a patient called Rhys Donnan. Donnan was in his 30s, had severe knee pain, and had been through the surgical mill. His first operation had been to correct a dislocated kneecap. Then he'd had several arthroscopies before undergoing a partial knee replacement, all of which culminated in his being crippled with arthritis and, by the age of 24, rendered nearly immobile. Seven years later, in chronic pain, Donnan limped into Barton's clinic.

'He was very depressed. He was suicidal, and he couldn't do anything without intense pain. I put a staircase as part of the assessment in front of him, and I said, "I want you to step up on this,"' remembers Barton. 'He said, "My knee hurts before I even start doing it."' Barton regrouped. He reassured Donnan that he wouldn't be damaging his knee, then started him off low, asking him to step up onto a single book that he'd pulled off the shelf. Donnan did it.

Over several days, Barton added another book, and another, gradually increasing the height. 'Within a week, he was pain-free doing those activities,' says Barton. Not satisfied with climbing a small flight of portable stairs, soon after Donnan scaled the Sydney Harbour Bridge with 'no problem at all' and went to Peru to trek the Inca Trail.

Donnan's story is extraordinary, but perhaps more astounding is the fact that Barton sees patients like him every week. How are they getting better? Barton says that such a staggering improvement, in just seven days, occurs too soon to be explained by exercise making muscles stronger. It's too soon for inflammation in the joint to have resolved. Critically, many patients are just like Donnan — they experience pain if they even think about moving their knee.

'In osteoarthritis, we have surgeons and general practitioners saying, "Look at your X-ray. This is all structurally changed. You now need to be careful, stop running, stop doing this exercise,"' says Barton. 'That induces fear. If they're fearing activities because they think they are going to be damaging and harmful, then their threshold for pain probably comes down.' Fear, from a belief that movement is dangerous, ramps up pain perception as part of the brain's attempt to protect the injured body part. Even worse, Barton thinks something in the knee is feeding the fear.

You might recall that, in central sensitisation, things hurt that normally wouldn't, something called allodynia. I experienced allodynia myself when just a gentle breeze on my knee would torment me. As Barton points out, the inflammation of osteoarthritis will raise the temperature and cause swelling and pressure, all of which alter sensations from the knee. Those sensations, sifted through the filter of fear, are ripe to be misinterpreted as pain. Fear can manufacture pain from sensations that are not, intrinsically, painful. Exercise can reduce that fear. 'We're telling people they're not going to damage their knees by walking up and down stairs. But we're also doing that task in a graded manner where we're starting on small steps, building to higher and higher steps,' says Barton. 'They're learning that these things shouldn't be harmful to their joints, and then they're practising it and testing it out for themselves.' It is, he says, an extension of education — a reinforcement through experience that movement is not a threat.

If changing mindset and reducing fear are so important, how does building muscle strength, as Lawrie did through GLA:D, actually help? Intuitively, you'd think stronger muscles would take load off the knee and reduce compressive forces through it, especially in someone who's overweight. But one recent study found that, while stronger quads did improve knee pain in osteoarthritis, it wasn't by reducing force through the knee or altering biomechanics when people walked. The researchers simply couldn't explain it. Barton, however, points to his own observations in the clinic. 'If you watch someone run who doesn't have good quadriceps strength, they'll often land with quite a straight knee,' he says. 'They're shunting all the forces into the internal structures of the knee. If you improve their quadriceps strength, they can take those loads with the knee bent when they hit the ground, which means they can absorb those loads through their quadriceps, rather than the internal structures of their knee.' This landing pattern, says Barton, changes sensations in the knee, which are picked up by mechanoreceptors and fed into the waiting brain. It's still a theory, but that altered load-bearing might reduce the sensations that the brain misreads as pain.

It's a picture complemented by a stark recent finding. Spanish orthopaedic surgeon Eduard Alentorn-Geli led a team that scoured studies about osteoarthritis in runners. They looked at 17 in total, containing interviews with more than 100,000 people. The team found that a little over 13 per cent of elite runners, some of whom were professional, had knee or hip arthritis. A control group of non-athletes with a sedentary lifestyle did slightly better: just over 10 per cent had arthritis. But when they ran numbers on recreational runners, whose mileage was decidedly lower than that of the elites, the incidence of osteoarthritis plunged, to just 3.5 per cent. When it comes to osteoarthritis, the study suggests, not moving at all is as bad as moving too much. 'If you're under-loading the joint, if you're sitting on the couch not doing exercise, your joint structure will deteriorate faster,' says Barton. One reason is that loading a joint prompts the creation of proteins in the cartilage lining — called proteoglycans — which keep it plumped up and resistant to all the pounding. Moderate exercise, the study suggests, limits osteoarthritis in the knee, and all the sensations that come with it.

It's now three and half years since Lawrie was scheduled to have that knee replacement. It still hasn't happened. He's walking five kilometres a day and keeping up with his GLA:D exercises. Yes, exercise hurt when Lawrie started, but the pain gradually wound back, in keeping with what Barton says is a common trajectory: pain typically improves around two to six weeks after beginning an exercise program. How did Lawrie get better? Perhaps exercise triggered endorphins or Sluka's anti-inflammatory effect. Or exercise-induced hypoalgesia was pushed along by the power of supportive words. Maybe Lawrie had less fear and a new set of beliefs about the benefits of exercise, so his brain simply didn't make pain when he moved the knee. Perhaps the altered biomechanics helped, too. Whatever the mechanism, it's great news because, while a knee replacement can give good pain relief, it doesn't work for roughly one in five people and, like all surgery, comes with potential complications.

All of which raises something baffling. The main reason to get a knee replacement is to relieve pain. But if your pain can get better without surgery, what does an operation actually do?

Chapter 5

Passengers on the Bus

questions about surgery

Roxboro Middle School is a red-brick pile with a looming chimney that runs its full three storeys before cutting a dramatic silhouette against the sky of Cleveland, Ohio. Beneath its west facade is a desultory car park, asphalt crisscrossed with the dribble lines of crack repair, bounded on one side by a rusting cyclone fence that wobbles to the batting cage of the school's baseball field, and on the other by a trio of tennis courts, their once-crisp green squares given a mottled wash by the weather of years. Whispers of grass poke through cracks in the tired surface. There is, in short, little to tell the casual observer that here once stood the Harold T. Clark Tennis Courts, a venue that hosted some of the game's greatest ever players.

On a blazing summer's day in August 1970, fans streamed into the main arena, a rectangular arrangement of bleachers — scaffolds topped with wooden benches that were marked, like rulers, with numbered white dashes for seats. A woman in Jackie O sunglasses and a sleeveless frock handed out programs for the event, the Davis Cup final between the US and West Germany, and patrons, men in short-sleeved white shirts and thin ties, women in Twiggy dresses, grabbed them eagerly. It was a doubles match with Stan Smith and Bob Lutz hitting out for the US team. Arthur Ashe was on the bill for a later match. One spectator had secured himself a prime viewing position courtside, but, curiously, the man, in his late 30s with short dark hair in a side part, didn't look like he was there for pleasure. He was holding a state-of-the-art Nikon Super Zoom-8 movie camera, which, as Stan Smith stretched up for a thundering serve or coiled his torso for a smashing backhand, the man pointed straight at the champion, who was soon to be ranked number one in the world. The man was a surgeon named Robert Nirschl, and he was there to solve a problem that had ended the game of tennis for an untold number of players.

A few years earlier, Nirschl had graduated from the orthopaedic program at the prestigious Mayo Clinic, in Rochester, Minnesota. Then he got a job at the Georgetown University Medical Center in Washington, DC and moved the family east. They settled in McLean, Virginia, an upmarket neighbourhood favoured by diplomats and government officials, a skip across the Potomac River from the capital. Residents of the blue-chip locale had recently banded together and bought seven acres of woodland to establish the Tuckahoe Recreation Club, carving out space for a swimming pool and four tennis courts. Nirschl, who lettered in tennis at high school, had let his fitness slide, so he signed the whole family up at Tuckahoe and was soon a regular on the courts, spending hours honing his game. Not long after, however, something put a stop to all that dedication. Whenever he hit the ball, especially a backhand, Nirschl would get a pain on the outside of his right elbow. A nagging pain that, even with rest and changes in his technique, just wouldn't go away. What was wrong with his elbow? That is something perhaps best understood with a simple experiment you can do on yourself.

Rest your forearm on a table, palm down, and run your other hand over the outside of the elbow. You should feel a bony prominence just up from the funny bone. From this jutting portion of the upper arm bone, three important muscles arise — I'm going to call them brevis, longus, and communis for short. They sprout as one before dividing and snaking down the top of your forearm to attach to the bones on the back of your hand. When these three muscles pull together, they do a critical job: they pull the wrist up. That movement is called extension, and so the muscles are known as extensors. In fact, if put your hand on that bony origin as you bend your wrist skywards, you can feel the tendons, which attach each muscle to the bone, moving. On a typical day, you might do this to comb your hair or get your hands into typing position, nothing too taxing.

On the tennis court, wrist extension can make or break players, whether they're once-a-week social hitters or Wimbledon champions. That's because wrist extension is critical for the single-handed backhand. Which is where all the trouble starts. With every cracking follow through, the backhand delivers a force of around 70 newtons, roughly equivalent to the weight of a bowling ball. Deliver that force again and again and eventually the things that attach the extensor muscles to the elbow — the extensor tendons — can get overloaded. And it hurts. Nirschl, after all those weekends at picturesque Tuckahoe, had given himself a case of tennis elbow.

As you might expect for a sports-medicine expert, Nirschl knew what was wrong immediately. But the more he thought about it, the more something didn't gel. Nirschl knew as well as anyone that one reason top players became world-beaters was the incredible amount of force they put on a ball. But if you scan the ranks at the top levels of the game, rates of tennis elbow don't go up. They actually drop. Nirschl was gripped with curiosity. Why do elite players dodge tennis elbow? If he could find out, he might be able to get on top of his own pain and, perhaps, solve a problem for his patients. Around four out of five people with tennis elbow got better within a year, but 20 per cent had ongoing pain and often had to give up the game altogether. Maybe Nirschl could help those folk, too?

He started scanning the literature. He read paper after paper, but not a single one satisfied him. Nirschl, now on the faculty at Georgetown University, decided he was going to have to do his own research. He would start by analysing the strokes of top players. He would need a movie camera, so he bought the Nikon. But then there was the question of access to players at the highest level. As luck would have it, the captain of the US Davis Cup team, a man called Donald Dell, happened to live nearby, in Washington, DC. So Nirschl set up a meeting and explained his dilemma, and Dell arranged for him to film the team. Hence Nirschl's prime spot in Cleveland on that hot August day.

Nirschl went through spool after spool on his super-8 camera. Then, back in McLean, he pored over the footage of the US players hitting forehands, backhands, volleys, and smashes. Soon, he began to see a pattern in their shots. But he needed images from another group of players for comparison: those at his own level; the amateurs. So Nirschl started touring the country clubs of Virginia, from Tuckahoe to the Washington Golf and Country Club, and beyond. He took hundreds of photos of men in short white shorts and women in short white skirts, in various phases of various strokes. There was no lack of enthusiasm, but there was, Nirschl noticed, a key deficiency in one area.

'The most important thing was the production of impact force,' says Nirschl. He is 87 now. His hair is a silvery grey but is still swept across in that neat side part, and his grey-green eyes peer out with the steely intensity of a man who remains sharp. He continues to see patients at the Nirschl Orthopaedic Center, in Arlington, but did his last operation in 2015. 'Inexperienced players, by and large, use their forearms for both control and force, or power. World-class players or experienced players use their forearms for control, but they use gravity and body-weight transfer and bigger muscles for power.' Those weekend hacks were, quite simply, using the forearm to do two jobs when it should have been doing one. The extensor tendons at the elbow were being pushed beyond their limit.

Nirschl was confident he'd made a big advance in

understanding how technique made amateurs prone to tennis elbow. He shot off an abstract to speak at the conference of the American Academy of Orthopaedic Surgeons, it was accepted, and, in the winter of 1973, Nirschl went to the meeting, along with thousands of attendees, in the cavernous interior of the Las Vegas Convention Center, a flying saucer–shaped monolith with a vast circular auditorium that had hosted the Beatles, Led Zeppelin, and Muhammad Ali. 'I was on a panel at the end of the day. I think they stuck me there because they said, "Well, what is this all about?" All the other talks had ended, so our little panel of three or four speakers were the only ones left. So we had the biggest crowd — there were maybe four or five thousand people.' Nirschl's presentation on stroke mechanics went down well. But there was a small group of doctors in the audience who paid very close attention. They were starting a new journal, and, after the talk finished, they approached Nirschl with a request. Would he write an article for its inaugural issue? Nirschl obliged, and the article, called 'Good Tennis — Good Medicine', was published in the *The Physician and Sportsmedicine* in 1973. Which is when people really started to notice Dr Robert P. Nirschl.

The journalist Jim Kaplan came and interviewed him for a profile in *Sports Illustrated*. Arthur Jones, inventor of the Nautilus exercise machine, got in touch, and Nirschl installed the equipment to condition his patients. Nirschl was fast becoming the go-to guy for people with tennis elbow, and they got Rolls-Royce treatment. Nirschl would analyse their technique. He gave them personalised exercise programs. For those needing extra help, he might inject steroid around the extensor tendons. But there were always people, no matter what Nirschl did, who just didn't get better. Inevitably, they would end up asking Nirschl the same question. Is there an operation that might help? 'I'd say, "Well, we've got this old-time thing, and, if you want to try it, let's give it a whirl,"' says Nirschl.

That old-time thing was called a Hohmann procedure, pioneered by the Bavarian orthopaedic surgeon Georg Hohmann and described in his 1927 article 'Über den Tennisellenbogen'. The rationale for the operation was simple: you had to do something to weaken the force put through the spot where those tendons attached to the bony outcrop next to the elbow, called the lateral epicondyle. What did that mean in practice? 'You just took a scalpel and found the tendon as it attached to the lateral epicondyle, and just cut the tendon,' says Nirschl. It sounds barbaric, and Nirschl wasn't happy doing it, not least because the procedure only worked about 50 per cent of the time, as often as a favourable coin toss. At no stage did the surgeon identify where the problem was, essentially making a heroic cut through soft tissue and hoping for the best.

Nirschl decided it was time to go back to basics. He headed into the cadaver lab at Georgetown to spend hours cutting into the yellow, preserved flesh of the cadavers' forearms, the acrid stench of formalin irritating his nostrils. He found communis and, just beneath it, longus. But there was another member of that trio of muscles that bent the wrist up: brevis. Where was it? Nirschl dissected deeper and found something troubling. The origin of the brevis tendon was hidden away under longus. Nirschl grasped the significance straightaway. Surgeons who did the Hohmann procedure never saw brevis at all. Could brevis, Nirschl wondered, be where the real problem lay?

He wasn't going to find out with more cadaver inspections — there were no ex-players on the dissection tables. The only option was real live patients with tennis elbow. 'If we're going to do an operation, then I think what we're going to have to do is, we're gonna have to move the extensor longus to uncover and examine the extensor brevis and see what happens,' Nirschl recalls thinking at the time. So he began discussing it with patients who were at the end of their tether. Nirschl remembers the exchange with the very first patient who was up for surgery. It was 1975. 'They basically said, "Hey, do the operation," and I said, "Well, I'm going to be straight off the reservation here, but I'm going to look at your extensor brevis and see what's there."'

To watch Nirschl operate is to marvel at a feat of human dexterity. When he makes that first incision through the skin next to the elbow, and peels it back to reveal the dull, off-white lustre of the tendon, his hand is steady, deliberate. He grasps longus with a pair of forceps and then makes deft, insistent passes at its underside with the scalpel blade to reveal the brevis tendon. On that very first foray, the root of the problem soon became apparent. 'There was not only obvious damage in the tendon, but there was a tear in it, and so it was quite clear that this was a pathological area,' says Nirschl.

He began offering the surgery to carefully selected patients who hadn't got better with standard rehab — about one in 20 of those referred to him with tennis elbow. He did operation after operation, and, each time, when he uncovered brevis, he saw the same thing. Instead of a firm white tendon, these people had a soft grey tendon, weeping with fluid. He sent samples off to the lab, where they were stained and examined under the microscope. What Nirschl had seen with his naked eye was even more dramatic when blown up.

Normal collagen forms orderly, parallel pink lines when it's stained, like quality wood grain. Nirschl's patients had tendons that looked like fibreboard, a bunch of cells sprayed together and loosely compacted. The collagen was completely disorganised. 'Like a Kansas wheatfield hit with a tornado,' is how Nirschl puts it. Cells called fibroblasts were pushing in, an unruly rabble, trying to bring small blood vessels with them, all of them deformed. This was not inflammation, Nirschl concluded, but degeneration. Semi-dead tissue. He started removing it from patients, with careful, delicate strokes of the scalpel. Why did he remove that tissue? He sums up the answer in a Nirschlism, one of the many aphorisms that he has pulled together as bite-sized guides to surgical practice: 'I view the failure here, in essence, as a heart attack of the tendon. It's unhealthy tissue. Unhealthy, generally, is pain-producing.'

By 1979, Nirschl had operated on enough patients to publish his results in a series of 88 surgeries. His success rate put the archaic Hohmann procedure to shame. Ninety-eight per cent improved, and 85 per cent returned to vigorous sport. Nirschl was now being sought out by the biggest names in tennis. He operated on Stan Smith, who'd slumped after winning Wimbledon in 1973. After the surgery, Smith went on to win the US Open doubles in 1980, edging out John

McEnroe and Peter Fleming. Nirschl worked his magic on the Spaniard Manuel Orantes, who, just 59 days after the procedure on his playing arm, beat Jimmy Connors for the 1977 US Men's Clay Court Championships. In appreciation, Orantes sent Nirschl the racquet he used, which takes pride of place on the wall of Nirschl's Arlington consulting room. Then there was Wimbledon champion Richard Krajicek, followed by countless others. In 1996, Nirschl featured on one of the top-rating US morning news programs, *Good Morning America*. In 2017, aged 84 and in the twilight of his storied career, Nirschl received one of the highest honours a doctor can get. The Mayo Clinic, consistently ranked the number-one hospital in the world, bestowed Nirschl with its distinguished alumni award, given to just a handful of outstanding medicos each year.

But as Nirschl received the award, among the Tiffany lamps and chaise longues of the library at the grand Mayo Foundation House in Minnesota, something unsettling was happening on the other side of the world. Two doctors were putting the finishing touches to a paper whose conclusion, given the pedigree of Nirschl's operation, was almost unthinkable: the Nirschl procedure didn't work.

Perhaps George Murrell was destined to ruffle feathers. His own father, Timothy Murrell, working as a doctor in Papua New Guinea in the 1960s, had overturned long-held theories about food poisoning linked to pig meat. He found the cause was a bacterium, not an allergen or virus as had been thought, which led to some consternation among those still invested in the old theories. In fact, it was while Murrell junior was showing slides of his father's groundbreaking research to his Year 11 biology class in Adelaide that his own medical ambition shifted front of mind. 'My class obviously saw something in me that they'd never seen before. It was just really cool, and the buzz I got from doing that pushed me towards wanting to do research, and wanting to do medicine,' says Murrell, who's now 61. Not only did he finish medicine, he got a Rhodes Scholarship to study abroad at Oxford University.

Then, in 1989, he went to the United States to do orthopaedic training, including a two-year stint in New York at the world-leading Hospital for Special Surgery, next to the East River in Upper East Side Manhattan. 'During the clinical year of that fellowship, I was exposed to a lot of sports orthopaedic operations, and one of the more common ones was this Nirschl procedure for tennis elbow,' says Murrell. 'Definitely the surgeons that I was working with were convinced it was a very useful procedure for managing recalcitrant tennis elbow.'

But a decade later, something happened that raised doubts for Murrell. He was at a conference in Washington and the speaker was a man named Bernard Morrey. Morrey was a world-renowned elbow surgeon and head of orthopaedics at the Mayo Clinic. It was a position with plenty of status, but it also meant Morrey was referred the hard cases no one else could fix, some of whom were people who had received the Nirschl procedure and didn't get better. 'He gave the impression that he didn't really think the surgery was that worthwhile. There were a lot of failures that he was seeing,' says Murrell. At the time of the conference, Murrell was once again living in Australia, working as an orthopaedic surgeon in Sydney, with a thriving private practice. He had also been made an associate professor at the University of New South Wales, where he was director of the Orthopaedic Research Institute. Given the uncertainties about the Nirschl procedure, he thought, why not do a study?

As it happened, he'd been approached by a young medical student named Martin Kroslak who was looking for a research project. Murrell pitched him the idea, it stuck, and the two set about designing the study. It would, they decided, be a trial that compared the real Nirschl procedure to a fake one. After giving proper informed consent, half the patients would get the real deal; the other half would get an operation that would follow all the steps of the Nirschl procedure, but stop at the point where the weeping grey tendon tissue would normally be removed. It would be a classic sham surgery study, designed to show if patients were really getting better because of that final surgical step.

The pair's first hurdle was to get their proposal through the ethics committee of the local health service. It flew though ethics, says Murrell, thanks largely to some rheumatologists on the committee who shared the desire for more evidence. But then the problems began. 'The public hospital wouldn't let me do the trial,' Murrell recalls. 'They said, "How can you do placebo surgery in a public hospital? Either your surgery works or it doesn't."' He was starting to get a feel for what his father had been through, butting up against the embedded orthodoxy of the establishment. So they presented it to the local private hospital and, much to their relief, got enthusiastic backing.

Shortly thereafter, in the autumn of 2005, ads appeared in the *St George and Sutherland Shire Leader*, calling for volunteers with chronic tennis elbow to take part in a study.

On a warm November day in a small hospital in the suburb of Kogarah, Murrell operated on the first volunteer. Kroslak had prepared some plain white envelopes, each containing a slip printed with either 'sham operation' or 'Nirschl procedure', and put them in a plastic folder. Before the patient was wheeled into theatre, while Murrell scrubbed and gowned at the stainless-steel trough in the adjoining room, a nurse pulled out one of the envelopes at random and slipped it in the patient's notes. Then the operation began. Murrell found the bony prominence beside the elbow, made his incision just above it, located communis, elevated it from longus, and then dissected the latter to reveal the origins of brevis. The tendon was dull and grey with the classic changes of tennis elbow. Now the anaesthetist opened the envelope, took out the slip of paper, and showed it to Murrell. It said 'sham operation'. What did Murrell do? He merely noted the appearance of the tendon, then closed the incision. That was it. He left that grey tissue — Nirschl calls it 'unhappy' — alone.

If their tussle with the public hospital had seemed hard going, it was nothing compared to what happened next. The pair laboured for a full decade to recruit enough volunteers to meet a statistical benchmark. The Nirschl procedure, says Murrell, just wasn't a common operation in these parts, perhaps because Australian doctors and physios didn't think it was useful. In April 2015, with operations completed on 26 people, 13 in each group, Kroslak and Murrell finally brought the study to a close. They had followed each person for two and a half years with a battery of tests. They had measured pain at rest and with movement. They had measured stiffness and tenderness. They had measured difficulty twisting and picking things up, the ability to bend and straighten the elbow and even — using a device Murrell helped design and build — the exact force with which each person could extend their wrist on the bad side. Murrell, however, never saw any patient again after operating on them. Kroslak did all the follow up and remained, just like his patients, unaware of who had got the real surgery and who had got the sham.

What did they find? On all the pain measures and most of the functional measures, patients who got the Nirschl procedure made major gains. But so did the people who got the fake procedure. They all got better. When Kroslak and Murrell crunched the numbers, there were, quite simply, no differences between the groups. Why not?

Murrell is tall and lean, with a full head of hair that is mostly still blond, and he's coping with the cool Sydney winter in a roomy tweed jacket over shirt and tie. He won athletic blues at Oxford — his personal best in the triple jump was over 15 metres — and he's kept in excellent shape. He's also surprisingly mild-mannered, given all that he's done, which is perhaps why he's able to admit, with no small amount of humility, to the frailties of the surgical endeavour. 'The human body is incredibly good at getting itself better, and we just, kind of, get in the way,' says Murrell. 'We either get in the way or we're passengers on the bus.' In other words: when doctors operate, they are often just along for the ride as diseases get better — via their natural history, because symptoms regress to the mean, or thanks to the placebo effect. Whatever the reason, the study's key finding was that removing the tissue Nirschl believed was pain-producing offered no advantage over leaving it alone. Why didn't it help? 'It's a bit Neanderthal, like a lot of things we do in orthopaedics, to think that cutting out a bit of macroscopically-looking bad tissue would solve the problem,' says Murrell. 'It's clearly an overuse-type injury. There's a mismatch between what the tendon can handle in terms of load and what load it's getting. To think that you can reset the whole thing by cutting out a bit of tissue, I think is a bit, now I know, it's a bit naive.'

Nirschl is a really charming guy. He's super smart, a virtuoso in the operating room, and has a CV to die for. He's helped countless people with a combination of exercise, medications, and surgery based on extensive research, much of which he carried out himself. He reckons he's done around 2,000 Nirschl procedures and, by and large, gets fantastic results. But what Murrell's study suggests is that the benefit from the operation is not, as Nirschl once put it, from scraping the 'old paint' off the tendon. The mere belief that you've had surgery and the expectation of getting better will, almost certainly, play a role. But there is something else, something very special about Robert Nirschl that might also explain why his patients do so well.

In 2011, a young psychologist named Luke Chang had a dilemma. Like all of his colleagues, he really wanted his patients to do better. Which meant he needed to become a better therapist. The standard way to do that was to become expert at the part of psychotherapy that really makes a difference. In cognitive therapy for depression, that might be getting the client to question whether their negative thoughts stand up to scrutiny. The therapist might challenge the client's belief that, for example, just because his girlfriend broke up with him, *no one* thinks he's attractive. Chang could simply drill harder into the mechanics of therapy to deliver it better. But then there was the small problem of his office.

Chang was doing an internship at the University of California, Los Angeles and had been installed in a room at the Semel Institute for Neuroscience and Human Behavior. The Semel was just a short Cadillac drive from the Santa Monica foothills and the fabulous mansions of Bel Air, and right opposite the spanking-new Ronald Reagan UCLA Medical Center. But it sat between them like an abandoned orphan. There was nothing either opulent or new about the Semel. 'It was, like, an old decrepit building. It had fluorescent lighting, the walls were just gross and stained,' says Chang. 'There was, like, this dirty linoleum.' On top of all that, there were cockroaches. So Chang got busy. He covered the lino with a rug. He put pictures up to hide the wall blemishes. He invested in a lamp to create a softer light and turned off the neon strip lights. Then he got to work on himself. He ditched his tennis shoes for some upmarket loafers. He dusted off his sports jacket. He put on a tie.

Why did he do all this? Chang knew that up to 70 per cent of the results of psychotherapy had nothing to do with the therapy itself. A lot of people get better because of something else altogether. Something called common factors. These are what happen in any therapy, regardless of whether it works on negative thoughts or unpacks your deepest Freudian neuroses. One common factor is a close, confiding relationship between client and therapist. Another is that the rationale for therapy, the basis on which it is thought to work, is accepted by both parties. But it was a different factor altogether that Chang was trying to exploit with his office tidy-up.

People who *expect* the therapy to work are more likely to get better. 'You are basically selling hope,' is how Chang puts it. At the time, he was a trainee and desperate to come across as the kind of experienced, caring, and authoritative type that his clients would have faith in. His sincere wish was that some interior decoration and a wardrobe makeover would convince his patients they were in for seriously good therapy. But for all his efforts, there was something destined to work against him, something ingrained in his persona.

Chang comes across as a wunderkind. He has thick black hair cut to a medium crew, dark eyebrows above intense, hazel-brown eyes, pale skin with not a line to be seen, and a dusting of beard that looks to have only recently been introduced to the razor. He's wearing steel-blue half-rim glasses, and sports some subtle neck bling above a dark open-necked sweater. He sprinkles his chat liberally with 'like' but, in between, shows an arresting breadth of knowledge about everything from psychology to computer programming and machine learning. It's not surprising, then, to learn that he's 41 and not 25. Ironically, the intellectual qualities that make Chang a whizz were thwarting his efforts, back in that dismal office at the Semel, to be a better therapist.

One really important common factor is allegiance. Therapists who truly believe their therapy will work, who pledge their allegiance to it, get better results. Chang, however, was an arch-sceptic. He was heavily of the view that therapies worked mostly because of things peripheral to their supposed mechanism. He wasn't a true believer in psychotherapy, and that was bad for his patients. Frustrated, Chang started thinking more about allegiance. He'd converted a dingy office into a half-decent therapy suite. He'd easily smartened up his attire. How could he conjure up the benefits of allegiance?

He would, he concluded, have to understand how the impassioned loyalty of therapists to their therapy rubbed off on clients. He would have to study allegiance, manipulate it, wind it up and down, and see what happened. But that meant giving some therapists faith in their therapy and others none — impossible in the consulting room, where each psychologist's bias towards a treatment was rusted on. Then Chang had an idea. What about pain? He knew that a patient's beliefs about painkillers could help pain through a placebo effect. Maybe, Chang wondered, the *doctor's* beliefs about a painkiller, their allegiance to it, could also influence pain?

The opportunity to test his ideas came a few years later, when Chang landed a job at Dartmouth College, where his postdoc mentor and legendary placebo researcher Tor Wager also worked. Chang put together a team that included Wager to study the problem. The first challenge was to control the doctor's belief in the effectiveness of a painkiller. That ruled out using real doctors, who are already wedded to their favourite drugs. Psychology students, they decided, would step into the doctors' shoes. They put the word out at Dartmouth, and nearly 200 signed up. One by one, as winter approached in 2015, they trooped in to the looming, historic Moore Hall on the Dartmouth campus in Hanover, New Hampshire, just a few hundred metres from the treed banks of the idyllic Connecticut River. There, they donned blue surgical scrubs and white coats like they were auditioning for the intern roles in *Grey's Anatomy* or *ER*. The casting notice, however, was a little left-field. They were told they'd be administering a painkiller to a patient, but, to understand how it worked, they'd first have to experience it themselves.

Chang escorted each 'doctor' to one of the labs on the second floor, sat them down at a small desk, and brought out a tube that had a label on it that read 'Thermedol. 7% Icapthenol HCI Jelly. TRP Blocker'. Then he squeezed some of the Thermedol onto a small dish and had the doctor mix it with red food colouring. Next, Chang took out a jar of Vaseline petroleum jelly, the age-old balm for dry skin. He squeezed some Vaseline into another dish and had the doctor mix it with blue food colouring.

Now Chang rubbed some of the red gel and some of the blue gel on different bits of the doctor's forearm. Then he brought out a thermode, the device that delivers a painful burning sensation. Chang pressed the hot thermode on the skin coated in blue Vaseline. Yes, the doctors said routinely, it hurt. Then he pressed the thermode on the skin coated with red Thermedol. As expected, the doctors said it hurt a lot less. The Thermedol, containing the TRP channel blocker, really worked. There was just one tiny problem. Check any drug guide — you won't find Icapthenol in it. And while TRP channel blockers are under investigation, none has yet made it as a bona fide painkiller. Thermedol is a complete fabrication. It's just Vaseline, too. So how was it stopping the pain? It wasn't. Chang was simply turning down the heat on the thermode whenever he pressed it on skin covered in the red Thermedol gel. The first part of his experiment was now complete. The doctors had pledged allegiance to Thermedol as a superior painkiller, oblivious to the fact that it was plain-old petroleum jelly.

It was time to bring in the 'patients'. Volunteers were dressed the part in a white patient gown, and took a seat opposite the doctor. The doctor Q-tipped some Vaseline onto each patient's arm and applied the hot thermode. Yes it hurt, was the universal reply from the patients. Now the doctor smeared Thermedol cream on the patient's arm and pressed the hot thermode there. It hurt a whole lot less. Thermedol worked. Yes, this time it *really did work*. Chang never turned down the temperature of the Thermode — it was set at 47 degrees Celsius (117 degrees Fahrenheit) all the time. But how on earth could petroleum jelly suddenly become a real painkiller?

Chang and his team had designed something ingenious to find out. It was a wrestling helmet with a rectangular frame of PVC pipe holding a GoPro camera that could record the helmet-wearer's face. The doctors had it on during the consultation, and an elaborate software set-up designed by Chang and his tech-savvy team registered their exact facial expression. When the doctors applied heat to the Vaseline-covered skin, something remarkable was happening. Ever so subtly, the doctors were displaying a wince of pain. But when the doctors applied heat to the Thermedol-covered skin, the face-reading software didn't pick up anything unusual. Their expression remained, implacably, neutral. The face, barometer of all that we feel and broadcaster of emotion to the outside world, was transferring the doctor's unassailable belief in the effectiveness of Thermedol to the patient. Thermedol was, and remains, off-the-shelf petroleum jelly, but the doctors had pledged allegiance to it. And it worked.

What does Chang's extraordinary study, published in *Nature Human Behaviour* in 2019, say about the Nirschl procedure? Maybe this: maybe Nirschl saw so many people get better after surgery, he became convinced it worked. Maybe he developed a profound allegiance to the operation as a boon for tennis elbow and passed that on to his patients, with the fleeting facial expressions that Chang discovered. But something else about Nirschl might have helped it all along.

There is a clue in that *Good Morning America* interview from 1996. Here's how veteran TV anchor Charlie Gibson introduced the segment: 'Today's subject is tennis. Playing tennis, a terrific way to stay fit. But what's great for your heart may be murder on your elbow,' said Gibson. 'With us now, orthopaedic surgeon Dr Robert Nirschl, a pioneer in the treatment of tennis elbow, and it's good to have you with us.' What did Nirschl say? 'Oh, it's our pleasure.' Our. It's *our* pleasure. I did a double take when I first heard this. Nirschl used the first-person plural, the 'royal we'. When you think about its origins, that's a little cringey.

On one account, the majestic plural dates back to 12th-century England and the reigns of Henry I and II, when it was invoked to indicate 'God and I' and the divine right of kings. 'We', when referring to the singular self, is a form of high-status branding. But it isn't reserved for royalty. In 1989, Margaret Thatcher famously announced that 'we have become a grandmother'. It's even more widespread than that. Pilots use 'we' and 'our' more often than their subordinate first officers and flight engineers.

But there's something even more curious about the royal we. In 2013, the psychologist James W. Pennebaker, from the University of Texas at Austin, led a study that examined group behaviour in person, the text of everyday emails, and the wording of a trove of correspondence between soldiers of varying ranks. Pennebaker found that higher status people used 'we' more. But they did something else as well. They used 'you' more. They were more 'other-oriented', a trait that might have got them into a leadership position in the first place, because it aligns with being fair, generous, and, crucially, more helpful. Which is why another of Chang's findings is so intriguing. Patients reported that doctors were more *empathetic* when using Thermedol, the treatment the doctors 'knew' would work. Chang can't be sure why, but he has some thoughts. 'If you know that the person is going to feel pain, you're feeling responsible and you're on edge, whereas, if you know it's going to work, you might just be calmer, and more soothing, or talking more, paying more attention.'

Here's the thing. In study after study, the same traits are found to predict a placebo response: the confidence of the doctor;

how positive they are about the likely outcome; and the degree of attention and empathy they show to the patient. Nirschl had each of those in spades, bound up, if we accept Pennebaker's findings, in a sense of status that brought with it a generosity of spirit to others. Dr Nirschl was, in sum, a prime candidate to transmit a placebo effect to his patients.

Which is not, it must be said, how Nirschl sees it. Kroslak and Murrell's study was, he says, deeply flawed. They only had 26 patients. They never sent any samples of the brevis tendon to the lab to confirm it was really damaged. 'We have no idea whether there was pathology or not,' he says. 'One of my better quotes to my fellows is that if you operate on a normal and don't screw it up too much, the results can be pretty good.' Nirschl says there was a fundamental problem with the sham surgery. When you cut through communis and longus to see brevis, you are, in effect, 'releasing' some of the muscles that pull on and overload the tendons. A kind of mini version of the Hohmann procedure. To get to that grey, semi-dead stuff on brevis, says Nirschl, you have to do something that also helps tennis elbow. It wasn't really a sham operation. But then, I asked Nirschl, isn't it impossible to do a sham Nirschl procedure? 'In essence, it is,' he says. Which makes the effectiveness of the Nirschl procedure, awkwardly, unfalsifiable. The Australian study won awards and was also, on a recent analysis, rated as excellent quality. Murrell accepts, however, that low patient numbers were a limitation.

It is a tall order to get any trial perfect. But as Kroslak and Murrell were drawing their study to a close, someone was trying to do just that. He was aiming to conduct the perfect sham study, whose conclusions would be beyond reproach, and he was doing it on something that just about everybody, as they get older, has wrong with them. In the process, that man would change the practice of medicine. Yet there would be a cost.

A few blocks in from the Ferry Building on San Francisco Bay, in the area known as SoMa, there is an enormous building with a low-slung, glass-cliff facade that looks like an airport terminal crossed with a shopping mall, spread over an area equal to 50 soccer pitches. The Moscone Center has had a colourful past. It featured in the 1995 Sandra Bullock thriller *The Net* and, more recently, hosted the Fortnite Battle Royale booth at a video-game conference. But something went down at the Moscone in 2001 that, arguably, would affect many more lives than either.

On a brisk day in February, a man named Bruce Moseley, a surgeon from Houston, gave a talk at the 68th annual meeting of the American Academy of Orthopaedic Surgeons. As Moseley spoke and flashed up slide after slide with graphs, facts, and figures, the enormous auditorium was silent, with just the occasional murmur rippling across the attendees seated in orderly rows. But when Moseley finished, all hell broke loose. 'People were furious. They crowded the microphones. They were completely out of their minds. They yelled. They almost cried,' remembers Teppo Järvinen, a Finnish professor of orthopaedics who, as a trainee surgeon, had been part of Moseley's rapt audience. 'We watched Bruce Moseley drop a bomb on the medical discipline.' What had Moseley done to stir a crowd of dour medicos into a churning frenzy? He had challenged a practice that was the heart and soul of orthopaedic surgery.

At the time, there was one operation that surgeons would commonly offer people with osteoarthritis of the knee. It was called an arthroscopic debridement and lavage. The surgeon would insert the metal arthroscope, inspect the inside of the knee, smooth over any damage to the cartilage that lines the joint, trim any tears of the meniscus, remove any debris, and flush it all out with a few litres of sterile salt water. Each year, 650,000 of the procedures were done in the US alone, at a cost of $3.25 billion.

Moseley, however, was dubious about its efficacy. So he put together a study of around 160 people with knee osteoarthritis, and divided them into three groups. The first group got the standard arthroscopy, with the smoothing, trimming, and washout. The second group got an arthroscopy with just the washout. The third group got placebo surgery: the surgeon made three incisions at the front of the knee, bent and straightened it like they do in the real operation, but never put the arthroscope into the knee.

Moseley followed the people for two years, checking their pain and how their knee was faring in activities like walking or climbing stairs. Why did the Moscone auditorium go ballistic? Moseley had found that people who got the placebo operation did just as well as those who got the full trim or the washout. This operation, the surgeons' lifeblood, which they had offered to patients time and time again, entirely in good faith, was a waste of time. The distress running through the Moscone Center was palpable.

Järvinen, on the other hand, was captivated. 'It was just a life-changing moment,' he says. 'Oh my god, what a beautiful trial, what truly practice-changing data.' Moseley's study had planted a seed. Järvinen headed back to Finland, knuckled down, and started his residency in orthopaedic surgery at the University Hospital in Tampere, a town famous for being girt by lakes and for its blood sausage, a delicacy known as mustamakkara. He hit his straps quickly, and it soon became obvious that one procedure would occupy a good deal of his time.

People, mostly older, would come in with knee pain, sometimes after an injury, sometimes out of the blue. Järvinen would examine them and order an MRI scan. Very often, he would find they had torn a meniscus, that C-shaped piece of shock-absorbing cartilage that sits in the knee joint. What happened next was a well-oiled routine. Järvinen would place them on the waitlist for day surgery. Then, when a spot opened up, he would do an operation to trim the torn meniscus and pare it back to a nice smooth edge — an arthroscopic partial meniscectomy, or APM for short. 'That was kind of the bread and butter of a young resident. A very simple procedure, easy to learn. You got to learn how to use the scope and how to manoeuvre around the knee.'

As Järvinen performed surgery after surgery, and became ever more skilled, an idea slowly took shape. He had wanted to do his own sham study since hearing Moseley in San Francisco. Why not, he wondered, do it on the APM? But Järvinen's motivations were not iconoclastic. He didn't want to show the APM was useless. On the contrary, he wanted to show how good it was. In his youthful exuberance, Järvinen would become the avenger of a profession that had just had one of its procedures thrown under a bus by Moseley's findings.

When you look at Järvinen's background, it's not hard to figure out why. He grew up in Turku, Finland's oldest city, which sits at the apex of the Archipelago Sea, a conglomerate of thousands of islands that splinter, shrapnel-like, amid the frigid waters of the Baltic Sea. His father, Markku Järvinen, was not only an orthopaedic surgeon, but also a pioneer of arthroscopic knee surgery in Finland and doctor to the Finnish Olympic teams for Montreal in 1976 and Moscow in 1980. Sporting celebrities were a fixture in their house when Järvinen was a little boy — among them, distance runner Lasse Virén, the Flying Finn, who won four Olympic gold medals. 'It was cool for a young, sports-enthusiastic kid. I was too afraid to talk to these stars — they were big heroes at the time. What impressed me was that my father would talk to them just like anyone else,' he remembers.

Järvinen approached his boss in Tampere, a veteran surgeon named Raine Sihvonen, and the two chewed over ideas for a study and began plotting their strategy. To prove that the APM worked, they would need to show that trimming the cartilage, and it alone, made people better. In the business, this is called an efficacy trial. So they made a checklist. Patients would have to be uniform: between 35 and 65 years old, with knee pain for at least three months. The type of tear would have to be uniform: only degenerative tears, ones that happen as the meniscus gets weaker with age. The site of the tear would have to be uniform:

medial tears, on the inner of the two cartilages. Why all the winnowing? Järvinen and Sihvonen wanted to ensure that, theoretically, the single tear was at the root of the person's pain. How would they be certain? Each tear would be confirmed on the gold-standard MRI scan. But they were still left with a problem.

Degenerative tears of the meniscus are common as people age, but so is something else. Osteoarthritis is almost the exclusive preserve of older people. The two happen together so often that a meniscal tear is now seen as part and parcel of osteoarthritis. The thing is, osteoarthritis can hurt, too. Järvinen and Sihvonen had to be sure that osteoarthritis wasn't causing their patients' pain, as that would muddy the waters. They resolved to take X-rays and prove each patient was free of osteoarthritis of the knee, but that was fraught as well. Radiologists looking at knee X-rays often disagree on the severity and even presence of osteoarthritis. So Järvinen and Sihvonen decided that a musculoskeletal radiologist and the senior surgeon at each hospital in the trial would have to agree that there were no, or minimal, signs of osteoarthritis on the knee X-rays of prospective patients. Then they recruited the best orthopaedic surgeons in Finland, who, between them, had thousands of hours experience doing the APM. They were ready to begin.

In 2007, in the depths of the Finnish winter, an orderly wheeled the first patient into theatre. The anaesthetist gave them a spinal anaesthetic and light sedation; they would be numb below the waist but awake, if a little drowsy. Sihvonen painted iodine on the bad knee, put a sterile drape over it, and made two cuts at the front. In one, he put the arthroscope. Sihvonen found the torn meniscus with the arthroscope's camera, and swivelled the TV monitor so the patient could see the damaged cartilage. The footage was recorded as proof. Then Sihvonen asked the nurse to turn the TV monitor back, so the patient couldn't see anything. The nurse opened an envelope, removed a piece of paper, and showed it to Sihvonen. It was the randomisation slip, printed with either 'resect the meniscus' or 'placebo surgery'. Sihvonen read it in silence.

What Sihvonen did next went against everything he, and every other orthopaedic surgeon, had been trained to do. He took out something that looks like an electric toothbrush, with a narrow metal barrel sticking out of one end. Inside the barrel, there is usually a tiny burr, like a miniature medieval mace, that rotates at high speed. Buzz it along the frayed edge of a torn meniscus and it eats through the friable cartilage, which gets sucked away by the arthroscope, leaving the meniscus clean and smooth. But today there was no burr in the barrel, and Sihvonen would not put the shaver inside the knee. The paper read 'placebo surgery'. He turned on the shaver and held it against the outside of the patient's knee, to mimic the sound and sensation of the real thing. Then he ran the pump that squirted saline through the arthroscope. It made loud slurping noises, but no torn cartilage was hoovered away. This was all an elaborate hoax with a singular aim: to make the patient think they were getting actual surgery. The trial, called FIDELITY, was officially underway, and Järvinen was confident it would show the APM to be a winning operation.

Meanwhile, 800 kilometres away and across the Baltic Sea, in the historic Swedish town of Lund, a young researcher named Martin Englund had been beavering away at another study. Englund, who also had an appointment in Massachusetts at the Boston University School of Medicine, was part of a research team that had tasked itself with something a little prosaic. Their job was to random-dial folk in Framingham, a little town on the outskirts of Boston. When someone answered the phone, the team member would ask them if they were aged between 50 and 90 and could walk. If the answer was 'yes', the researcher made an odd request. Would they agree to have an MRI scan of their right knee? It was tedious, but, in the end, nearly a thousand people obliged. What the researchers wanted to find out was this: if you take a sample of older people, off the street, how many will have a torn meniscus? The answer, they found — and as I reported in the introduction — is nearly a third. Just over 308 of those Bostonians had a torn meniscus. But Englund and the team were also interested in something else. What did it *feel* like to have a torn meniscus? It turned out that, for most of those people, 61 per cent to be exact, it didn't feel like anything. They had no symptoms at all. No pain, no aching, no stiffness. Intrigued, the researchers started scrutinising their data with another key question. Did any of this change if you had osteoarthritis? So they zoomed in on people who had at least slight changes of osteoarthritis on their X-ray and were also getting pain in the knee: 63 per cent had a torn meniscus. On the traditional view — that a torn meniscus hurts — the finding made a lot of sense. But then the team looked at people with osteoarthritis who had *no knee pain at all*. How many of them had a torn meniscus? Bizarrely, it was almost exactly the same proportion: 60 per cent.

Having a torn meniscus, Englund had shown, was totally compatible with having no pain at all. The discovery let off a depth charge under Järvinen's study. It struck at the whole rationale for APM surgery. If there was no concrete link between a torn meniscus and pain, how could there be a concrete link between trimming the meniscus and getting rid of pain? But it was even worse for Järvinen. Although the results of Englund's study came out in *The New England Journal of Medicine* in September 2008, he had actually pointed out the disconnect between a torn meniscus and knee pain in an article back in 2004, when Järvinen was just beginning to design FIDELITY. Järvinen hadn't read it.

'We designed FIDELITY as an efficacy trial, to become the most popular guys in our discipline,' says Järvinen. 'We realised, now that we actually knew what we should have known when we were designing the trial, it was highly unlikely that it was going to work.' Järvinen was blindsided, and embarrassed, but he pushed on. It was a slog. Because meniscal tears were so often seen with osteoarthritis, it was almost impossible to find people with just a tear and no arthritis. But when they did recruit someone, the protocol was rigorous. Järvinen himself never saw or operated on anybody. He was totally at arm's length so he couldn't 'contaminate the data' with the kind of transmitted placebo effect that Luke Chang had discovered. The surgeon who operated never saw the patient again, either. Each patient was followed up by a separate team, none of whom knew which person had got what. For a full year, Järvinen's researchers tracked their volunteer patients, with everybody in the dark on who got the bogus surgery. At two, six, and 12 months, the researchers asked each patient to fill out a questionnaire. Did they have pain in the knee after exercise? Was it locking up so they couldn't bend or straighten it? Did they have a limp? How much pain did they have overall? Could they squat? Could they climb stairs?

In March 2012, nearly five years after beginning the study, surgeons had operated on 146 people, half of whom had received the sham surgery, which was enough to start analysing the results. But Järvinen's obsessive rigour still wasn't satisfied. He set up a writing committee to do something you don't see very often. They split the data so they had a set of results for each of the two groups, those who had received the APM and those who hadn't. But they didn't know which was which. They called the first group 'A' and the second group 'B'. First, they assumed group A got the real surgery. They drew their conclusions, and each author in the writing committee signed off on them. Then, they assumed group B got the real surgery. They drew their conclusions again, and again each author signed off on them. They had, effectively, locked themselves in to what they believed the study showed. There would be no biases even in the way the results were written up. After nearly half a decade of hard graft, only then were the blinders taken off.

It's nine in the morning when I chat to Järvinen. It's fair to say he's not at his best. He was up writing into the night, has only slept four hours or so, and he's groggy. Pictures from the early days show a striking man, square-jawed with a clipped beard and thick, swept-back strawberry-blond hair. A Norse warrior in dark-rimmed spectacles. He's filled out a little since then, the warrior's face is puffy, and there are dark circles under his eyes. He's well groomed, however, his beard still neat and his hair combed. He's wearing a fresh polo shirt in lime green and white, and his skin has been burnished to a golden glow by the Finnish sun. As he tells me what happened after his study was published, in the same top-tier journal as Englund's, I can't help but wonder about the toll it's taken on him.

The study, of course, found that the people who got the real surgery and the fake surgery did equally well. On average, each group improved a lot. Trimming the meniscus did nothing to improve pain or function. Which was a problem for the most frequent procedure done by orthopaedic surgeons. In 2006, 425,000 were done in the US alone. It was also a problem for Järvinen, who can remember exactly when things started to go downhill for him.

'You have to understand that my dad was a sports orthopaedic surgeon, so I knew most of the guys, the big names, the inventors of the arthroscopy. Many of them had visited our summer house in Finland, and they were good friends,' says Järvinen. 'I used to visit the arthroscopy conferences from two to five times a year. I would see these guys, have a cup of coffee. We would discuss how their families were doing, how they had become grandfathers. I think, over the course of 2013, I met one of the big guys five times. And then everything changed. I can tell you the day exactly because it was Boxing Day 2013.' What was so special about Thursday 26 December 2013? It was the day Järvinen's study was published. 'The next time I saw those guys, they didn't even look at me. I tried to say "hi", I raised my hand, they walked by like they didn't know me.'

Järvinen had produced a gold-plated study and was now a pariah in arthroscopy circles. It wasn't long before the criticisms started appearing in journals. A group of surgeons claimed *The New England Journal of Medicine* was biased against arthroscopic surgery. They argued that no person 'in his or her right mind' would consent to sham surgery and so the results were not 'generalisable to mentally healthy patients'. Another critique opined that, 'unfortunately, any study investigating [a surgery] on individuals born and raised in Finland cannot be extrapolated to the rest of the world' because Finns embrace *sisu*, a special form of courage, grit, and determination that doesn't exist elsewhere, affecting how they deal with pain. What does Järvinen say? His participants gave careful, well-considered, informed consent. And if the inner strength of *sisu* influenced the results, it did so equally in both groups — they were all Finnish — so the key finding of no difference remained. Järvinen thinks the barbs smack of desperation: 'We ran out of bullets, now we are throwing rocks.'

APM rates remain high in the US, but they are falling in Finland, where, says Järvinen, many surgeons were involved in FIDELITY and saw the results firsthand. They also work mostly in the public system, so there are few financial incentives to keep doing the procedure. I ask Järvinen if he is confident that APM surgery will eventually be given up. 'Science advances one funeral at a time,' he says. It is a quote, drawn from the writings of the Nobel Prize–winning German physicist Max Planck, with a blunt message. Opponents are not won over with rational truths; rather, their shibboleths must die with them. Only when today's youth, fertilised in the rich earth of new ideas, become tomorrow's science leaders can there be a conceptual changing of the guard. Sadly, Järvinen's experience bears this out. In the last decade, he has not been invited to speak at a single international orthopaedic conference.

These days, George Murrell is a shoulder surgeon. A lot of people come to see him with one type of problem. If you lift something too heavy, too often, or too awkwardly, especially as you get older, you can tear the tendons that cross the shoulder. Surgeons often try to fix it with an operation called a rotator-cuff repair, a surgery that has been Murrell's daily bread for a good part of his professional life. Recently, however, studies have suggested that surgery doesn't improve pain or function any more than simply strengthening the shoulder with physiotherapy. Those studies will soon culminate in their own, inexorable climax. The rotator-cuff repair is about to be tested in a sham surgery study, led by Professor Ian Harris at the University of Sydney. It is called the Australian Rotator Cuff Study, and Teppo Järvinen is an associate investigator.

The study is very close to home for Murrell. He is nervous — worried it will show the surgery isn't helpful. What would he do, I asked, if Harris's team produced an impeccable study that showed no benefit of surgery? Would he give up doing the rotator-cuff repair? 'It would be hard for me, because I'm emotionally wedded to trying to fix rotator-cuff tendons that are torn,' says Murrell.

The waves of new science wash through, while the breakwater of what we already do holds fast. What should patients do? Harris, who is an orthopaedic surgeon and has written a wonderful book called *Surgery, The Ultimate Placebo*, has a simple answer: 'Just ask the surgeon, "Am I going to be better with the surgery compared to not having the surgery?"'

It is a question that is critical for people who have surgery to treat pain. And it is a question of evidence. You might have noticed something about the study by Kroslak and Murrell on the Nirschl procedure, and the study by Järvinen's team on the APM. In each case, one of the main measures of success they used was pain. Now ask yourself these questions. Why hasn't anyone done a sham surgery study to see if putting a plaster cast on a broken arm works? Or to see if removing a cataract from someone's eye restores their sight? Or to see if doing a liver transplant for someone dying from liver failure can save their life? Because a healed bone, restored sight, and normal liver function are all plain to see. They are objective. You can measure them with an X-ray, an eye chart, a blood test. Pain is different. It is a chameleon. It might come from damaged tissue. But it might just as well come from sensitised nerves. Or it might be learned, through conditioning. And pain relief could well, as Luke Chang showed so beautifully, depend on something as capricious as how invested a doctor is in the treatment. The way they look at you. How much they care for you.

When pain is involved, chances are higher that the doctor is just 'a passenger on the bus'. In which case, it would be a mistake to dive in and fix the anatomy with surgery. If a surgeon says most of their patients get better after surgery, that doesn't tell you how they would do without surgery. And if the surgery works via a placebo effect, wouldn't it be better to get that without having surgery? Because surgery is never without risk. Järvinen recently published the results of the five-year follow-up on FIDELITY participants. Both groups maintained similar improvements, but a worrying difference emerged. The group that got real surgery had a 13 per cent higher rate of osteoarthritis in the operated knee. Not only does APM offer no advantage, FIDELITY finds, but it might even be harming people.

Sham studies do not show that there is no place for surgery. What they do is give us information, the bedrock for informed consent. Without it, how can people make up their minds about whether the surgery is a good thing for them? Let's face it — in the end, that decision always belongs to us. The very least we can ask is that all the information is on the table when we make it.

Chapter 6

The Safety Matrix

the role of hypnosis in treating pain

On a limpid summer day, just after Christmas 2018, Anne Zeestraten and her nine-year-old daughter Koraly were soaking up the holiday sun near the pier on Pompano Beach, one of many lengths of warm, beige sand overlooked by high-rises along the coast of the Miami metropolitan area of Florida. Koraly had just come out of the water after a swim, happened to look down, and noticed something unusual. A lump had appeared at the top of her right leg, just below the line of her bathers. She showed her mum straightaway. 'I thought maybe she got bitten by something in the water,' says Zeestraten. 'She wasn't sick at all, and she didn't have any fever.' Nonetheless, Zeestraten thought it wise to get it checked out, so they headed to the emergency room at a nearby hospital. The doctor did some tests and then disappeared for what seemed an inordinate amount of time to check out a simple bite. 'It wasn't normal.

Then he came to see me and said, "You know what, you should go back home and talk to your doctor."' He handed Zeestraten, who is fluent in French and her native Dutch, and gets by in English, a referral letter. 'They gave me a letter in English that I didn't really understand because it was, like, doctor things.'

Zeestraten was starting to get worried. They flew home to Montreal and went to the GP. It was mid-January, and the Canadian city had been hammered by its worst blizzards in a century. The snow was knee-deep, cars lined the streets like over-iced gingerbread houses, and the temperature plunged to well below freezing. The GP referred Koraly to the Montreal Children's Hospital, but it would be a glacial wait before they could be seen. 'In March, the hospital called us, and we went there on a Friday,' says Zeestraten. 'By the next Monday, we knew. So it was very, very fast after that.' The hospital had scheduled Koraly to have the groin lump removed, under a general anaesthetic, and the biopsy confirmed their suspicions. She had a rare type of cancer called childhood Hodgkin lymphoma, which causes a type of white blood cell to grow uncontrollably.

The news was devastating, but there was little time for Zeestraten to take it in. Chemotherapy had to begin as soon as possible, but, before it could start, Koraly needed to undergo a short procedure. Chemo would take months and, to avoid having a new IV placed for every injection, the doctors wanted to put in a PICC line, a peripherally inserted central catheter. It's a plastic tube that goes in a vein at the front of the elbow and gets threaded all the way to the big vein that leads to the heart. Once it's in, it is there for the duration, and the chemo drugs can be given safely through it. Once it's in. The thing is, especially in children, it's not always plain sailing.

Montreal Children's Hospital is part of a sprawling medical complex, newly built on a flat expanse at the southern end of the island of Montreal, between the melting pot bustle of the Côte-des-Neiges neighbourhood and the lochs of the 19th-century Lachine Canal, a stone's throw from a labyrinthine flyover on Route 136. The feel is barren and industrial, but its designers have lightened the mood with multicoloured windowpanes on the entry facade, and, round back, there's some public art: a silver bear balances on a giant ball, and, further along, a big red heart, composed of crisscrossing straps that let the light shine through, sits atop a heavy grey plinth. On the plinth is an inscription, a line from a song by French crooner Charles Trenet: *Un bon sourire et tout s'éclaire*. It means, 'One good smile and everything lights up.' When you meet Vicky Fortin, you could be forgiven for thinking the line was written about her.

Fortin is 37, has dark hair in a bob of finger-wave curls, blue eyes bordered with smile crinkles, and a tinkling laugh that reminds you of childhood. When we speak, she's in elegant mufti with a woven white halter top. She tells me she started work as a medical imaging technologist at the Children's in 2004, straight out of college, and has been there ever since. During that time, she has seen just about everything that can go right, and wrong, when it comes to procedures in sick kids. Her stomping ground is a cavernous, neon-lit room on the third floor. It's filled with cabinets on wheels that contain drawer after drawer of medical equipment. There are hazardous-waste hampers, oxygen outlets, and, in the middle of it all, a solitary examination table with a large video monitor hovering behind and the ominous, robotic arch of an X-ray machine poised at one end. They have tried hard to cheer things up: a picture of a Pippi Longstocking–style girl with fluoro hair graces one wall, a solitary, stylised chicken alongside her; and when Fortin is at the helm, she wears a lead apron emblazoned with butterflies and a scrub cap that features tropical fish swimming through coral. The mood, however — especially when the machines lurch into action with a grinding whir and urgent beeps — is workmanlike, grave, and has a disturbing air of finality. It is not conducive to happy children.

Fortin has assisted in liver and kidney biopsies where a large needle is inserted into the child's abdomen, bladder studies that involve pushing a catheter into the urethra, nutrition support with feeding tubes that go into the stomach, and procedures where wires are passed into the heart. And she's worked with PICC lines, which she is trained to insert herself. Local anaesthetic is always injected to numb the area, but some children, understandably, still object. Often strenuously. In such cases, Fortin has a pre-agreed strategy with parents. 'We will always explain, outside the room, before we enter, what's the plan, what we will do as immobilisation, so that the parents know what's coming, and we would explain it to the child, of course, too,' says Fortin. 'Then we ask the family to let us know if at some point they feel like this is too much. If they feel that their child is not in a good place and that they want to abort the procedure. We wrap the child with blankets and then we use velcro to attach them to the table. Someone would hold the arm, you know, all the way up. We might play some videos and try to entertain them, but, anyway, some of them would scream, and some of the procedures were cancelled.'

When that happened, things got complicated. Fortin now had a scared and very distressed child who, understandably, had just been relieved of the modicum of trust they had in their medico. Worse, they were scheduled for sedation, with heavy-duty drugs such as chloral hydrate, midazolam, and propofol, which meant waiting for an anaesthetist, and the attendant risks of being sent off to sleep. All of which is without mentioning the deep unease staff felt about holding children down. None had any idea that there might be a better way. Until they got a tip-off.

In 2018, Fortin's then boss, chief medical imaging technologist Johanne L'Écuyer, heard about something curious happening at two hospitals in France. Doctors at the Rouen University Hospital, north-west of Paris, had been doing a variety of procedures, including varicose-vein removal and colonoscopies, normally done under sedation or general anaesthetic, without drugs. Instead, they had opted for something that many people rate, at best, as a cheap stage trick or, at worst, as snake oil. They were using hypnosis.

L'Écuyer, a chic, eloquent, no-nonsense woman, was intensely sceptical but, at the urging of a senior colleague, agreed to visit France to see if hypnosis was really something they could add to their toolkit. She and a colleague set out on a scoping mission to Rouen. They also went south, to the Hôpital Femme Mère Enfant, a paediatric hospital in Lyons that was using hypnosis for PICC lines and a procedure called sclerosis to remove skin haemangiomas, or strawberry birthmarks. To say the visit led to a recalibration of their world view would be an understatement.

'We were completely stunned,' L'Écuyer said when they returned. Hypnosis looked to have genuine effects. The technologist was now bent on running a trial of hypnosis at the hospital, but a big hurdle stood resolutely in her way. Montreal Children's Hospital is part of the McGill University Health Centre, the largest hospital system in Canada by bed capacity. McGill University itself is one of the world's premier universities, ranked 27 in the world as of June 2021. Could these august institutions possibly countenance the use of hypnosis? And in children?

L'Écuyer presented her case to the hospital executive, and impressed. They gave the go-ahead for a pilot study, and things moved rapidly from there. In January 2019, an expert hypnotherapist from the Lyons hospital, Claire Benoit-Ruby, was flown out to train four staff members. Shortly thereafter, L'Écuyer's team began to formally offer hypnosis to children having a range of procedures.

On a crisp spring morning in 2019, a shy, softly spoken, nine-year-old girl with a missing front tooth came to the medical imaging department with her mum and sat outside. Koraly was a little nervous, and so was Zeestraten. Soon, they were greeted by Fortin, who explained the procedure, that there would be a needle, and what a local anaesthetic was. But then she suggested something unexpected, and a little magical. 'Would you like to fly somewhere? Would you like to have a dream?' she asked Koraly.

This was the first Zeestraten and Koraly had heard about hypnosis. They didn't hesitate. Fortin took Koraly to the examination room, where they stopped at the door. She asked Koraly to look around and listen to the sounds of the machine, and they walked across to the bed. 'This is where you will lie down. It is all warmed up for you. I will put a blanket on you, and I will swaddle you like a precious gift. We will make sure you are very safe,' said Fortin. Koraly lay on the bed, and Fortin asked her to pick any object in view and focus on it. 'I want you to pay attention to the sound of my voice, the vibration of it. The more you will look at the object, the more you might feel that your eyes are getting heavier and blinking.' Now Fortin's voice changed, subtly, to a slow, rhythmic coaxing that wafted over Koraly like the rise and fall of a sea breeze. 'The more you put all your mind to it, the more you are sliding inside yourself. Just looking and listening. It is like there are bubbles around you. You can see bubbles in the air, their roundness, their colour, feel them inside your nose, sliding down your throat to your chest. Maybe just listen to the sounds of your breathing, in and out, and the more you breathe, the more you exhale, the more you blink, the more you slide inside. And at some point, your eyes will just want to close and rest, and you can let them open or close, as you wish. While you breathe, you may feel your legs getting heavier. Feel them on the bed. Feel the difference between the heaviness of your legs, and the light feeling of your chest breathing. Just like this, very, very easily, you will now to start to imagine.'

Then Fortin took Koraly to another place. She was in a room

alone. There was a set of stairs, leading down. Koraly walked down the stairs and came to a road. A ship was docked next to it, and Koraly climbed aboard. The ship sailed away, and it wasn't long before a big butterfly landed on the deck just next to her. Koraly clambered onto the butterfly's back, and they flew off together, swooping over the water. There were dolphins, whales, all kinds of fish. Then a bird appeared. It landed on Koraly's arm. She could feel the bird's feet on her arm, grabbing her skin. It wasn't comfortable. But it wasn't painful.

While Koraly was on her cruise and flight, under the harsh glare of an examination light, the radiologist had set up the PICC line kit, swabbed her arm, and laid out a sterile drape. That bird, scrabbling around with its feet for a good grip on Koraly's arm, was the local-anaesthetic injection and the PICC line being inserted through the skin at the front of her elbow. Did it hurt? 'Yes, it hurt a little bit,' says Koraly, who's now 11, 'but not very much.'

It's astonishing. Fortin had utilised the two elements of hypnosis — induction, where relaxation and breathing techniques help the person enter a trance-like state, and suggestion, the introduction of novel thoughts or ideas — to perform a procedure with local anaesthetic that, according to one recent study, requires sedation in nearly two-thirds of cases. How could the sound of a person's voice and an imaginary ride on a butterfly across the sea take the place of a powerful anaesthetic drug?

'You are saturating your brain with information, and your brain has so much information to listen to that you will forget everything that is around, and it will centre itself inside,' says

Fortin, her accent tinged with the singsong of Québécois French. The aim, she says, is to redirect attention from the outside world, from the 'critical awareness' of all the potentially worrying things in a medical procedure room, to an inner world of 'imaginary consciousness'. In this new world, things make their own sense. 'The story has to work with what the pain induces at that moment. If the bird is coming with its legs, this could be the poking moment. So the body will feel pain, but, if the pain makes sense with the situation you are in, you will just acknowledge it, like, "Okay, the bird is on my arm. It's not the procedure, it's not painful, it's this bird that came here."'

Fortin listens closely to her small patient, to their likes and dislikes, before the session. That way she knows if the destination is better to be an amusement park or a hockey rink, and the sensation on their arm the sun or the ice. The refocusing comes with a dollop of something called dissociation. When the sun shines strongly, it is always on 'the' arm, not 'your' arm. 'Just by a small technique like this, it is not *your* body, it is *the* body, you are dissociating yourself from this state *here*. So you can go somewhere else. It is like you detach a body part so you don't feel it.' And language is critical for another reason.

To understand why, it's worth trying a thought experiment that has entered the hall of fame in psychology circles, one whose lineage traces back to the great Russian writer Fyodor Dostoyevsky. In the summer of 1862, Dostoyevsky undertook a tour of Western Europe, documenting his often caustic observations in an essay titled 'Winter Notes on Summer Impressions'. During the trip, which presumably involved plenty of inner musings, Dostoyevsky became aware of one of the brain's recurring quirks. 'Try to pose for yourself this task: not to think of a polar bear, and you will see that the cursed thing will come to mind every minute,' he writes. Take a few moments to resist a white bear invading your own mental space and perhaps you'll share Dostoyevsky's intuition. The challenge was taken up formally more than a century later by US psychologist Dan Wegner, who noted in a series of experiments that, once introduced, ejecting images of a white bear from one's mind is well-nigh impossible.

The experiments have lessons for pain. As experts in medical language Allan Cyna and Elvira Lang point out, 'Negating words … fail to mitigate the effects of negatively valued words.' In plain terms, if your doctor mentions the word 'pain', even if they preface it with, 'don't worry, there won't be any', pain will hang around in your brain like a rogue polar bear. It is a lesson that has bred new language at the Montreal Children's Hospital. Phrases like 'I won't hurt you' and 'Don't be scared' are out, in the knowledge that children will hear 'hurt' and 'scared' and disregard everything else. Phrases that are in: 'I will take care of you' and 'You are safe'. When the sun comes out and shines on the child's skin, timed with the stinging local anaesthetic, there will be no 'burning' but, instead, 'Soon you will feel a sun inside your skin. The more you feel it, the more you are protected.'

In the months that followed insertion of the PICC line, Koraly had her chemotherapy, her hair fell out, and her mum buzzed what was left to a short crew. Koraly covered it up periodically with a baseball cap or a grey beanie with a big H at the front, the logo of her cherished ice-hockey team, the Montreal Canadiens, known to the locals as the Habs, for Les Habitants. Summer came and went, the silver maples on the streets by the St Lawrence River promenade littered the flagstones with an autumnal yellow carpet, and Koraly got some good news from her doctors. She was in remission. There would, in short order, be joyous celebration. But something else needed to happen first. The PICC line had to be removed.

This is usually a simple procedure, done by a nurse on the cancer ward. But when a plastic tube has been embedded in a blood vessel for months, the body sometimes has a way of not letting go, probably from adhesions forming between the tube and the blood-vessel wall. 'They tried and they tried, and they could not remove the PICC line because it was stuck,' Zeestraten remembers. 'They were pulling and pulling, and Koraly was in pain. She was crying, and we had to have a break because it was painful. They tried again, but it was still stuck. So it's why after they decided to do the hypnosis.' They took Koraly back to the procedure room. But the PICC line wasn't just stuck. It had fractured, leaving part of it adrift in the big vein near Koraly's heart. Now the radiologist had to do a new procedure, to access the vein from Koraly's other arm and snare the errant catheter tip. This time, however, all was calm, because Koraly, under the soothing tutelage of Fortin's whispered words, was elsewhere.

By November 2019, L'Écuyer's team had enough data to present the results of their pilot. In all, they did 80 procedures under hypnosis, in children ranging from five to 17 years old, including 48 PICC line insertions, five bladder scans, and 16 biopsies of thyroids, kidneys, and lymph nodes. After each study, they asked the child to point to how they felt on a pain scale of coloured faces, zero being a green happy face and ten a red face with a sad mouth and waterfall tears. In a control group of 28 children who'd had procedures without hypnosis, average pain was 5.4 out of ten. In the kids who got hypnosis, the average pain score was just 1.4 out of ten. It was only a pilot, and the results weren't formally published, but L'Ecuyer and her team were ecstatic, as was the hospital. It was time, they decided, to share the results with the public.

Koraly and her mum agreed to do a demonstration for the media, and a presser was sent out. Vicky Fortin would do the hypnotherapy. The media scrum duly assembled in the procedure room on a chilly day in December 2019. Among them was a cameraman from the hospital's comms department, who wore earphones that allowed him to hear Fortin, who was miked, as she talked Koraly into that hypnotic otherworld. In retrospect, given what happened next, that may have been a mistake. 'He was like, "Oh my god, my knees, my knees are falling under myself." He had to shake himself out of the feeling,' says Fortin. 'He was actually falling under the hypnosis just by listening to what I was saying to the girl.' It is a recognised side effect, she says, of the sotto voce trips to Disneyland and the beach that were now heard on a regular basis in the procedure room. 'Even the doctors who I work with have to be careful not to listen too carefully to me.'

It is powerful stuff, and a technique of increasing interest.

But it is held back by a paucity of research, perhaps fuelled by the raised eyebrows that the idea of hypnosis still elicits. Of the existing studies, some disappoint. A recent Australian trial of hypnosis for people having burns dressed — one of the most painful procedures you can have — found hypnosis didn't help the pain, although it did reduce anxiety and heart rate. On the flipside, a consensus published in the *European Journal of Cancer* in 2020 issued a 'strong recommendation' for 'the use of hypnosis for all needle procedures' in children with cancer, based on moderate quality evidence — four studies of 120 children.

Debates over efficacy aside, a big question remains for people with chronic pain. The type of hypnosis used at the Children's, known as Ericksonian hypnosis after its creator, the American psychiatrist Milton Erickson, targets pain in the here and now. Fortin had to be at the bedside, murmuring her dreams into the willing ears of her young patients. People with chronic pain will not, in general, have the luxury of a personal hypnotherapist. What they will want to know is whether visiting a hypnotherapist today can fix their pain tomorrow. If it can, a large subset will also be interested in something else. How?

Steve Olson grew up in Rockford, Illinois, an industrial town that straddles the Rock River in Winnebago County at the northern edge of the state. In the '60s, Chrysler opened an assembly plant in Rockford, and Olson's dad, an engineer, got work with a company that designed and built metal cutters and other machinery used in car manufacturing. Olson's passion, however, was always wood. The meandering grain of a polished tabletop or the stepped-down curves of a well-turned leg were his thing, and so his future was in the furniture business. He did repairs and finishing, and made legs for tables and chairs at Ethan Allen, one of America's biggest furniture companies. After 17 years, he moved on, designing and setting up franchise stores for a bespoke woodcraft company, eventually settling in Franklin, a small town on the outskirts of Nashville, Tennessee, known for the Victorian-revival architecture of its downtown precinct.

Life was good. But in 2009 the wobbles began. In the wake of the global financial crisis, the woodwork company took some big losses. 'After being there for about 20 years, they just called me up and said, "You no longer have a job." They were closing their franchise department,' says Olson. He fell into a depression but eventually regrouped, securing a job with a sports supply store in Franklin. It wasn't wood, but Olson was content with the daily chores of adjusting inventory on the computer, stocking shelves and hooks, dealing with customer queries, and climbing ladders to put stock away.

One day in October 2013, Olson turned up for work at the store, a massive barn of a building in a commercial corridor just north of the town centre, and the empty shelves happened to be ones down low. 'I was kneeling on the floor stocking shelves and had trouble standing back up,' remembers Olson. The thing stopping Olson getting upright was pain. 'It started with one leg and then went to the other. The pain was so intense I couldn't walk, I couldn't kneel down, I couldn't stand. About four inches above my knees, up to the middle of my thighs, it would feel like somebody hit me with a baseball bat. Multiple times, both legs.' Olson couldn't drive, so his wife picked him up and took him to the local emergency department. 'The doctor in ER gave me as much morphine as I could have to try and stop the pain, and it did nothing to help.'

Worse, they had no idea what was causing it. Olson got a referral to Tennessee's number-one hospital, the Vanderbilt University Medical Center, in Nashville, where he began months of tests and treatments to try and find out what was wrong and how to stop the pain. Among the gamut of investigations was something called electromyography: the doctor gave tiny electrical shocks to the nerves in Olson's legs to measure how quickly they could relay messages. The results were worrying. Olson's nerves were underperforming in a major way.

A few years earlier, Olson had been diagnosed with type 2 diabetes, which plays havoc with blood-sugar levels, but it can also do something else. The disease attacks small blood vessels, including the ones that ferry oxygen to the body's network of nerves. When that happens, the nerves get damaged, a condition called diabetic neuropathy, and it hurts. For Olson, the pain was unrelenting, especially at night. 'There were nights where I would just sit in a chair and cry for three or four hours because the pain was so intense. I couldn't sleep with a blanket covering my legs or even a sheet. It just felt like somebody was laying a big board across my legs and standing on it.'

Efforts to control the pain were doing precious little. 'The neurologists and the doctors I saw were just stumped. They couldn't figure out what to do for me.' He had amassed an impressive daily cocktail of drugs for the pain, including tranquilisers, muscle relaxants, oxycodone, sleeping pills, extended-release morphine, and the neuropathic pain drug gabapentin. Olson shows me a picture on his phone. It's a one-gallon ziplock bag and, inside, it looks like the tide came in and dumped a rainbow of flotsam — tiny discs and rods in yellows, pinks, greens, and oranges. It is, in fact, a snapshot of his diet of pills at the time. 'I was basically a vegetable. I couldn't really function.'

Olson's life was plunged into misery and disarray. He had to quit his job at the sports store. Desperate for a solution, he secured an appointment with a neurologist at the prestigious Mayo Clinic. 'He said the damage the neuropathy has done to my legs is forever and that I'll suffer with that pain the rest of my life.' In Nashville, the news was no better. 'The doctors back here told me that, in a few months, I'd probably be in a wheelchair for the rest of my life.' By now, Olson had tried just about everything — anaesthetic injections into his spinal nerves, epidurals, electrotherapy to pass low-level current through the skin, even injections of platelet-rich plasma from his own blood into the thigh muscles. He was considered for a spinal stimulator, a pacemaker-like device that sends electrical impulses to the spinal cord, but says his insurance didn't cover it. Nothing had worked.

After all those failures, it was, perhaps, to be expected that Olson soon began formulating his own treatment plan. 'The pain was so absolutely, incredibly intense, I just got to the point I felt like I couldn't deal with it anymore. It would be better, if it would stop the pain, to have no legs. I pleaded with the doctors to cut my legs off, but they wouldn't, they wouldn't listen to me. They said, "Unfortunately, it is not going to stop the pain even if we did do that."'

Olson is 70 now. He's a solid guy with receding grey hair and large dark-rimmed glasses above heavy cheeks, and he's wearing a no-fuss grey T-shirt. He punctuates our chat with a warm smile and easy laugh, but if his demeanour gives little away, that's because everything is in his voice. It is low-pitched, seldom varying from an andante beat, sometimes dropping to a croak, and resonating often with a deep timbre of sadness. His is a hard story to listen to.

The next thing the doctors tried him on was testosterone, levels of which can drop in people taking opioid painkillers. Testosterone replacement sometimes helps with pain, but Olson thinks it pushed him over the edge. 'I broke down. I didn't even know why I was breaking down and asking for help. I just sat on the back porch crying. I couldn't stop crying because I was in so much pain,' Olson was now having suicidal thoughts. He was admitted to the psychiatric ward with severe depression. It was October 2016, and he had been living with crippling pain for three years.

Olson was discharged from hospital after a week, and re-entered a world that had, he believed, nothing left to offer. Until someone happened to mention that Vanderbilt had a pain service Olson had never been told about. It was called the Osher Center for Integrative Medicine, and one of the services it provided was medical hypnotherapy. 'I was very sceptical. I didn't know if I believed in it at all,' says Olson. His sole contact with hypnosis had been those tragicomic stage shows on telly. 'They teach you to go out and cluck like a chicken, or you can't pick up a sponge off the stage because you think it's a brick.' Nonetheless, he decided to give it a try.

On an overcast day in November 2016, Olson fronted the Osher Center, a modern building wrapped in *Space Odyssey*–white panels that admits visitors through a zip-like crevice in its facade, located on Nashville's busy West End Avenue. Inside, he was greeted by a softly spoken woman with shoulder-length brown hair. Her name was Lindsey McKernan, and she was a clinical psychologist and associate professor in the Department of Psychiatry and Behavioral Sciences at Vanderbilt.

'The first session was to get to know her and talk a little bit about what I was dealing with,' Olson remembers. 'Then she basically had me find something in her office to stare at, whether it was a plant on a shelf, or a book, or a dot on the wall, and I just started thinking about that spot that I'm trying to pay attention to. Then she slowly started talking me down into kind of a relaxing, almost like a sleep. I just closed my eyes. I just felt very relaxed, very comfortable.' Soon Olson found himself on a path, then walking down steps, ten in all, to a door at the bottom, beside which was a box, in which he could leave all his problems. He opened the door and walked through it to find a bench where he could sit and, at McKernan's suggestion, experience less pain.

The session was interesting, but not earth-shattering, and Olson hobbled out of McKernan's office on the same cane he'd hobbled in on. Olson came back the next week and the one after that and, each time, the routine was repeated. On his fourth visit, however, something changed. This time, after he entered the relaxed state, McKernan had him imagine a control dial on his leg, like the volume knob on a radio. The dial, he soon discovered, had powers that imaginary things really shouldn't have. 'She told me, "If we turn it up, it will increase the pain, and, if we turn it down, it will decrease the pain." So we turned it up, and I instantly broke into tears and got shaky. It was absolutely off-the-chart painful. She immediately said, "Let's turn the volume back down, turn the pain back down," and I turned it down,' says Olson. What followed is seared in his memory. 'I walked out of her office without a cane.'

When he'd gone in, Olson's pain was a nine out of ten. Coming out from that session, freed from the cane he'd used for years, it was a two. It was an inflection point in Olson's life, heralding a new phase he could never have an anticipated. McKernan recorded some of the sessions, and Olson stored them on his phone. If he was out and about in Franklin with his wife and the pain came back, he would head back to the car, take a seat, and, to the soundtrack of McKernan's voice, drift off to the bench by the door, or adjust the dial on his leg and nudge the pain down. He's done the same on a plane trip — same result.

Medical hypnosis also prompted Olson to have a bit of a clean out. There was a reason he showed me that bag of pills on his phone. He's thrown it out, along with three other gallon bags stuffed full of tablets. Olson is walking three to five miles a day, he hasn't been back to the psychiatrist, and, while his pain levels bob around, they are way down. According to his formal case study, published by McKernan and her colleagues in the *American Journal of Clinical Hypnosis* in 2020, he's had a full two-point reduction on a ten-point pain scale, now maintained for five years since that first hypnosis session.

One of Olson's pastimes is catch-and-release fishing. There are a few ponds around Franklin, including the sandy-shored, clover-shaped one in The Park at Harlinsdale Farm. Old barns and disused silos clutter the pond's edge, and its muddy waters are stocked with large-mouthed bass. Olson couldn't fish for years. He's taken it up again, and there's a photo of him leaning forward, holding a whopping bass, its ample belly gleaming silver with camera flash, eager to help the photographer catch it for posterity. The fish must weigh five kilos. It's a moment that was unthinkable for Olson just a few years ago, when he couldn't even stand in front of a sink and do the dishes.

It is, undoubtedly, a remarkable recovery, but all the usual caveats apply. Could it be a placebo effect? Was Olson going to improve anyway? Might we be victim to the logical fallacy known as *post hoc ergo propter hoc*? Because pain relief that follows hypnosis doesn't prove it was caused by hypnosis. A recent analysis of 85 studies on experimentally induced pain found that hypnosis, especially in people who scored higher on ratings of hypnotic suggestibility, did lead to 'clinically meaningful reductions in pain'. It also noted the need for more studies in 'non-laboratory' pain.

Doubts notwithstanding, what could possibly explain an imaginary volume knob on your leg turning down neuropathic pain, among the most excruciating a person can experience? And there's something else extraordinary about Olson's recovery. For extended periods, his pain was under control while he went about his everyday business, with no need to enter a hypnotic trance. How could hypnosis one day influence pain the next?

A bit over a decade ago, when Carolane Desmarteaux was 19 years old, she was mucking around with her younger sister, who was 18, in their duplex apartment in the suburbs of Montreal. Desmarteaux wanted to be an actress, and her sister was into singing and dancing, so their sibling play had a strong performative edge. Desmarteaux was doing her best to imitate the stage hypnotists she'd seen on TV. 'You will relax, close your eyes and go deeper and deeper,' she intoned. On cue, her sister's eyes closed, and she settled back in her chair. Encouraged by her sister's method acting, Desmarteaux pulled out something else she'd seen in the stage hypnotist's bag of tricks. 'You will obey my commands, and you won't be able to do otherwise. When you open your eyes, you will be five years old.'

Which was when something strange happened. Her sister slowly opened her eyes and, at the same time, moved her legs back under the chair, hooking each foot around one of the chair's front legs, toes pointing outwards. 'That was the moment I realised she wasn't joking,' says Desmarteaux. It was a movement, she says, the family hadn't spoken about, or kept photographs of, but it was, unmistakably, what her sister always did as a little girl. 'I hadn't seen it for 15 years.' As a young adult, her sister was an extrovert, but at that instant, says Desmarteaux, she had become suddenly shy, her eyes glued to the floor, just as she was as a small child.

It's called age regression, and it's a big-ticket item in hypnosis circles; it's one of the measures on the Stanford Hypnotic Susceptibility Scale that, along with tests of whether your eyes close, your arm drifts, or your taste buds detect the sour acidity of lemon at the therapist's suggestion, gauges how easily you can be hypnotised. This dramatic demonstration got Desmarteaux very interested. Her first instincts were that hypnosis could be an excellent tool for one of her theatrical hopes — to be a director.

'Actors on stage, this is what they do, they generate alternative sensations to normal. They do believe it. They are in this alternative state, and they try to authentically connect with this feeling,' says Desmarteaux. Critically, she adds, actors enter that state through the vehicle of words — those of the playwright. Mind made up, Desmarteaux enrolled in a hypnosis school, the École de formation professionnelle en hypnose du Québec, near the docks of Montreal's old port in one of the 19th-century industrial buildings the government had rescued from dereliction. She graduated with flying colours and opened a practice in an unassuming brown-brick building on the Rue Masson, still with the goal of using hypnosis as a theatrical tool.

But as she began to see patients, many of whom had chronic pain, something changed. Desmarteaux would perform a standard induction and then, using the control knob imagery that Steve Olson found so helpful, ask her clients to turn the pain up and down. Mindful that people might want to reactivate that state at a later time, sometimes in situations where listening to a recording wasn't practical, she added something else: 'You give a key that he or she can use later and recreate what they have done in hypnosis.' That key could be something as simple as a gesture, like holding one's thumb and forefinger together in a circle, that revivified the relaxed state of hypnosis and brought the controls back within her client's grasp.

The result was a complete reversal of the antiquated understanding of hypnosis as something that the therapist does to the client. 'The key is an example of how you can gradually give back control to the person. At first, they might say, "Okay, it's the hypnotherapist, it's not me. There is something going on, and I don't control it." But gradually, they claim those tools, to take control again, to say, "Okay, I can act on the pain when it is too much. I don't have to submit to it. It is not something happening *to* me."' Her clients' new-found autonomy was a revelation to Desmarteaux, and she was gripped with a new urge. She needed to found out how it was working.

As it happened, there was a world expert just down the road. Pierre Rainville, an intense man with rockstar dark locks and a philosopher's mien, was a professor at the University of Montreal who'd been doing leading-edge research on hypnosis and pain for nearly two decades. In 2017, Desmarteaux signed up to do a bachelor's degree in neuropsychology and, two weeks in, knocked on Rainville's door to say 'hi'. Two years later, she was enrolled in a doctorate, with Rainville as her advisor, joining a team that was wrangling data from a study he'd begun in 2014. The study was designed to answer precisely the question

Desmarteaux was interested in: how could hypnosis one day influence pain the next?

On Queen Mary Road, across the street from Montreal's highest structure — the looming basilica of Saint Joseph's Oratory, with its bulging army-green Renaissance dome and phantasmagorical Art Deco interior — is a stern brown-brick building, shrouded in trees. The building, erected in 1932, was designed by the architect Alphonse Piché, who is eulogised in one biographical entry as a competent practitioner with the caveat that, among his oeuvre, 'not a single work stands out as being innovative or progressive'. Piche's building now houses the neuro-imaging unit of the University of Montreal's institute of geriatrics, whose work is about as far from that unflattering epithet as you can imagine.

In 2014, 24 people trooped into the building for a hypnosis study. They were mostly in their late 20s and had been screened to ensure none had chronic pain or was on pain-relieving medication. One by one, each person lay down in the MRI scanner. A researcher attached two electrodes to the skin behind their ankle, just over the sural nerve, which relays sensations from the foot and lower leg. Very soon, the researcher informed them, they would receive some moderately painful shocks to the ankle. Then the researcher put earphones on the person, secured a Darth Vader–style frame around their head, and asked them if they could see the little screen attached to it. After the person nodded assent, a cross appeared on the screen, and a voice through the earphones asked them to stare at it.

'Now that you are comfortable, relax. Relax completely,

deeply,' said the voice. 'Relax every muscle in your body. You are now going to have a surprising experience. Let your body become soft, soft, soft.' The induction was complete; now came the suggestion. 'Imagine that your ankle is becoming more and more sensitive. Your skin is becoming very sensitive, and you will feel the shock more and more, as if a layer of metal was placed between your skin and the shock. Now imagine that your whole foot is turning into metal. Remember that metal is a very good conductor.' Then the researcher triggered nine painful shocks, one after the other, and asked the person to rate how much they hurt on a scale of zero to 100. After a brief pause, the mesmeric voice started up again. 'Imagine now that your ankle is becoming less and less sensitive. Your skin becomes numb and you will hardly feel any shock. It's like a layer of rubber between your skin and the shock. Take the time to imagine your foot completely made of rubber.' Nine more shocks ensued, the scanner whirring all the while, gathering images of each person's brain in pain.

When we speak, Desmarteaux and the team have just finished their analysis of the data. It is an extraordinary study. The first finding that stands out is perhaps predictable: participants who were more hypnotically suggestible, on the Stanford scale, felt more pain after the 'metal ankle' prompt and less after the 'rubber ankle' prompt than participants who were less suggestible. The revelation is what happened in the brains of those 'high suggestibles'. People who modulated the pain more effectively, up or down, had greater activity in an area called the parahippocampal gyrus (PHG), which is a tongue of grey matter in the brain's emotional system.

'This region is linked to the integration of semantic information, not the wording as much as the sense,' says Desmarteaux. 'And the integration of the words generates more context.' It sounds a little cryptic, but Desmarteaux is describing a chain of events with profound implications for people in pain. Statements like 'your whole foot is turning into metal' and 'your foot is completely made of rubber' are, after all, just words — 'semantic information'. How do you turn a sequence of words into the altered perception of a body part? First, you need to retrieve the *sense* of the soft, spongy, insulating properties of rubber, and the hard, smooth, electrical-conducting traits of metal. Then you need to frame that novel sense within a *context*: here, the setting of a painful shock to the leg. The result is a shift in beliefs about what will be experienced when the shock finally comes. A similar thing occurs, says Desmarteaux, with the term 'white coat'. These are mere words that lead to the perception of a garment, whose meaning is transformed when it is draped on the shoulders of a person behind a desk with a stethoscope around their neck. That humble garment, in the context of a medical consultation, can raise blood pressure all by itself, something known as the white-coat syndrome, or convey a doctor's authority and enhance the placebo response to a pill.

The crux of the study is this: people with more PHG activity were better able to use words to change their beliefs and expectations about what would happen when they felt pain. The result was a shift in the balance of power between beliefs and incoming pain information. 'People will have the bottom-up information, the nociception, and the beliefs. The beliefs will have more impact. They are able to add weight to these beliefs,' says Desmarteaux. Crucially, these weightier beliefs are formed in advance of the painful stimulus. It is a mechanism, the team has found, by which hypnosis can act over time to reduce pain. There is more. The team explains the finding within the framework of predictive processing — that is, the power of expectations to modify incoming pain signals, such as we saw in Chapter 3. Hypnosis, for people who are suggestible, can imprint benign expectations that bat away the insistent, doom-like patter of pain signals coming from the body, a little like the muffling of rain drops on freshly tempered glass.

Desmarteaux loves neuropsychology, but she never gave up the theatre — in fact, she's doing a master's degree in theatre at the University of Quebec. She has managed to combine the two. In Quebec City, there is a neat brick building with a woodwork facade, painted vermillion, and the words 'Editions du Grand Midi' in stand-out gold letters above its shopfront French doors. By day, it is a publishing house, but, on a rainy evening in May 2019, five actors welcomed their audience to a play there. The lead was Desmarteaux, dressed in a white coat, and the four others, all women, were a musician, a mime, a dancer, and an actress. The scene was the Salpêtrière Hospital in Paris in the 1880s, where the great neurologist Charcot was delivering one of his famous Tuesday Lessons. He was using hypnosis to bring on, and ostensibly cure, the symptoms of 'hysteria' — contortions, outcries, fits, and delirium — in women under his care. The plaintive, haunting tone of a singer is backdrop to the almost-somnambulist stage moves of the actors. The audience is left wondering, are they acting or truly hypnotised? The answer, says Desmarteaux, who plays Charcot, is that they are really hypnotised. But there is a twist. She worked with the actors over three years to bring the production to life, and, during that time, she taught them self-hypnosis. It is a subversion of the authoritarian trope of hypnosis as a Svengali-like exercise in control, instead offering hypnosis as an exercise in autonomy and empowerment. The women, not Charcot, were the authors of their own hypnotic trance.

Some have called hypnosis a non-deceptive placebo. And the suggestions of hypnosis do sound an awful lot like a placebo. There is, however, at least one crucial difference. The pain relief you get from a placebo is helped along by endorphins — the body's natural opioids. It makes sense, then, that you can reverse placebo analgesia with the drug naloxone, which blocks opioids binding to receptors. But naloxone does not reverse the pain relief produced by hypnosis. Hypnotic analgesia does not, it seems, work through the body's opioids. There is something else going on.

At the University of Washington (UW) campus in Seattle, there is a circular body of water that, from above, looks like the darkened screen of an idle smart watch. Each autumn, however, the Drumheller Fountain, known to the cognoscenti as the 'frosh pond', springs into life as an initiation site for the university's freshmen, or 'froshes', who are ritually baptised with an obligatory dunking. Wander down from the fountain, and across Northeast Pacific Street and you'll find a hotchpotch of low-rise buildings overlooking the steel-grey waters of Seattle's Portage Bay. This is the UW Medical Center, consistently ranked as one of the world's top hospitals. Back in 1989, the hospital welcomed one of its own new recruits to the team at the multidisciplinary pain centre.

Mark Jensen had finished his doctorate in clinical psychology at Arizona State University and returned to Seattle to take up the role of attending psychologist. Jensen's daily fare would be a steady stream of people, all referred to him with chronic pain. Like any newbie, of course, there was a settling in period, but Jensen's practice soon fell into a well-oiled routine. His training had been in cognitive behaviour therapy (CBT), whose central tenet is to seek out cognitive distortions and put them right. Someone with lower back pain, for example, might believe the area is damaged beyond repair and that movement can only make it worse. Should the medical evidence rule that out, Jensen would have the person challenge those thoughts, whittling away at them until, eventually, they drifted off into the ether.

It was an approach that worked well, for a time. Early on, most of his patients had pain from their muscles and joints, often from injuries. But then Jensen began to get referrals for a very different kind of problem. He was seeing patients who, like Steve Olson, had neuropathic pain, coming directly from damaged nerves. These new patients were a major challenge for Jensen, for one key reason. 'Rarely did they have cognitive distortions,' Jensen recalls. 'They just hurt.' He was all at sea. In the absence of those skewed beliefs, whose correction was the central goal of his favoured CBT, Jensen had almost nothing to offer. He was, effectively, rendered useless.

It was deflating, and Jensen grew increasingly frustrated at his impotence in the face of neuropathic pain. He began looking for alternatives and, in the process, came across a book. It was called *A Whole New Life*, by an American man of letters named Reynolds Price. Price was a Rhodes Scholar, a friend of the poet W.H. Auden, and a professor of English literature who had been struck down with debilitating pain from a spinal-cord tumour in his early 50s. After suffering for years, Price finally found something that helped him turn it around — hypnosis. The story made a deep impression on Jensen: 'I thought to myself, "I probably ought to learn to do this."' He did, and, to his surprise and delight, hypnosis worked for many of his patients with neuropathic pain. Jensen had now expanded his armoury of therapies into a neat dichotomy: for his patients with cognitive distortions, he had CBT; for those without, he could use hypnosis.

But Jensen was still left with a problem. Although CBT worked, it had a drawback. When Jensen reviewed the increasing number of studies on CBT, its effects on pain were, overall, modest at best. For someone touting CBT as one of their signature treatments, that was an issue. Jensen mulled over the problem, to no avail. Then, in the northern spring of 1995, a study quietly appeared in the *Journal of Consulting and Clinical Psychology* that was to influence Jensen's thinking for the rest of his career.

The article was lead-authored by a man named Irving Kirsch,

then a psychologist at the University of Connecticut. Kirsch had analysed 18 studies, all of which had done something novel. They had combined CBT with hypnosis to treat an illness. The aim? To see if the two therapies together were better than either alone. There were studies on anxiety, snake phobia, and insomnia, but, as Jensen turned the pages, it became apparent that most of the studies, nearly half, had focused on a single big-ticket health issue — they had looked at whether the double treatment could influence the Holy Grail for many people: weight loss. What Kirsch found stopped Jensen in his tracks. CBT undoubtedly helped with obesity, leading to weight loss of roughly five kilograms. But adding hypnosis to CBT effectively doubled the weight loss, to around ten kilograms, and kept it off for at least two years. It was a stunning revelation. And buried in Kirsch's article was a study that spoke directly to Jensen's problem. Two researchers had added a hypnotic induction to CBT for people in chronic pain. Though it was a small study, the results were promising — but there the research trail fizzled out. It had, however, given Jensen an idea.

Up to this point, most hypnotherapists used direct suggestions to quell pain. 'There is a control dial on your leg — turn it down and you'll experience less pain,' for example. But Jensen knew that almost all patients, including those with neuropathic pain, were prone to catastrophic thinking. 'It's terrible, and I think it's never going to get better,' or, 'I wonder whether something serious may happen,' and the like. Oftentimes, those predictions were downright wrong. Jensen's idea was this: if he could target catastrophic thoughts with CBT and give it a boost with hypnosis, maybe he'd get better results. And maybe he could redeem CBT as a therapy for neuropathic pain.

At the time, there was one group of people that regularly knocked on Jensen's door for help. Multiple sclerosis attacks the myelin sheaths that insulate nerve cells, degrading their ability to send messages, which drop from a clear signal to a patchy sputter, like an out-of-tune radio. A miserable by-product of that nerve damage is neuropathic pain. Many of Jensen's MS patients were also plagued with catastrophic thoughts. They were, in short, an ideal group to decide if his radical new approach could hold water.

So Jensen put together a team, and they recruited 15 people for a pilot study. Each person got four different treatments. They got education, much of it about how the body and brain create pain. They got CBT, aimed at challenging those catastrophic thoughts. They got training in hypnosis, including a special cue to do self-hypnosis at home — 'take a deep, refreshing breath and hold it for a moment and let it go' — and suggestions to damp down pain. And they got CBT again, but under the sway of hypnosis. What did they find?

To understand, you need a quick handle on something called an effect size. It's a standardised measure of the change produced by a treatment — in this case, a reduction in pain. An effect size of 0.2 is considered small, 0.5 is medium, and more than 0.8 is large. Jensen's study, published in 2011, found education alone did very little to reduce the average pain intensity of those people with MS. CBT did somewhat better, with an effect size of 0.23. Hypnosis alone, however, was more than twice as effective as CBT — the effect size was 0.56. And the effect size of hypnosis combined with CBT? 'To our surprise, it was rather large,' says Jensen, with a hint of understatement. The effect size of hypnotic cognitive therapy was an impressive 0.96, nearly double that of hypnosis alone.

Jensen knew they were onto something. But he wasn't off to the races, and for good reason. Not only was the study a small pilot, but it had failed to address two key questions. The first would need answering if the treatment was to have any hope of being taken up by pain practitioners: how did hypnotic cognitive therapy work? The second question was one that almost every patient was going to ask: will it work for me?

In the cold of February 2013, Jensen was hunkered down at the UW Medical Center, having put together a new team and secured funding for a bigger study. They got busy. Over the next four years, they posted flyers in the pain clinic, scoured the university's medical system for people with persistent pain, and mailed out nearly 2,500 recruitment brochures, telephoning as many people as they could to follow up. It was thankless work, and there were a lot of knockbacks, but they managed to gain the cooperation of 173 folk with chronic lower-back pain or pain from MS, a spinal-cord injury, or a limb amputation. These folk were, on average, 54 years of age with an average pain intensity of nearly five out of ten.

This time Jensen's team assigned them to get just one of those four treatments — education, CBT, hypnosis, or hypnotic cognitive therapy, following them up for 12 months. In 2020, the team published their findings. They were emphatic. All four treatments, including education, which had seemed ineffectual in the pilot, led to large reductions in pain intensity — effect sizes ranging from 0.6 to 0.8 — maintained for 12 months. But again, hypnotic cognitive therapy came up trumps, the only one to be statistically superior to education alone, which was the baseline. Nearly 60 per cent in this group said their pain was much or very much improved. Adding hypnosis to CBT really did help pain.

Jensen is 63, bespectacled, with a full beard and a crumple of receding hair, all of it tinged with a black that is giving way to grey. His voice pushes along at a canter, rising and falling as one or other idea sets off an impassioned explanation. His presence is paternal, infectiously optimistic, and saturated with dedication to his field. One focus of the recent study, he says, was to make 'helpful' thoughts more automatic. Take a person with lower-back pain where the evidence has ruled out ongoing damage. 'You say, "Those sensations that you experience are normal, and some days they will be more and some days they'll be less. But what is important to understand is that your back is strong. It is safe. It is durable."' When the therapist introduces those helpful thoughts, says Jensen, you want the client's 'record button' to be 'on'. How then, do you flick that switch?

On a typical day in the life of the brain, different groups of neurons pulse along at different speeds, a bit like heartbeats that quicken and slow. These are our famous brainwaves. There are groups of neurons that fire at faster rates and are really active when you're wide awake and focused: those faster beats are called beta and gamma waves. Then there are groups of neurons that fire at slower rates and are more active when people are feeling relaxed, in a state of wakeful rest: those slower beats are called alpha and theta waves. 'It appears to be the case that slow-wave oscillations are incredibly important to new learning,' says Jensen. 'What happens when you get into a hypnotic trance, what happens when you say, "Listen to my voice, focus on my voice," is that the brain decreases activity overall, but the number of slow waves increase. What we think is that when you help somebody get their brain into a state where there are more slow oscillations going on, the brain is ready. The record button is on.'

Jensen stresses this is speculation, but his team did something else in the study, part of which is being prepared for publication, that speaks to the issue. They were trying to pin down the folk likely to do better with hypnotic cognitive therapy. They ran everyone through the Stanford Hypnotic Susceptibility Scale and found, not surprisingly, that those who scored higher did better. But they also measured something else. Before each treatment, they hooked everyone up to an EEG device to measures their brainwaves. Some people had more alpha-wave activity; it was simply the way they were wired. Those people also did better with hypnotic cognitive therapy. The finding fit with the idea that alpha waves set people up for more effective brain change. A surfeit of slow waves was, in short, priming the brain's signature means of executing new learning: neuroplasticity.

Jensen's team also looked at how people's thought processes changed. Catastrophic thinking, they found, decreased in all four treatment groups, and this decrease was linked to improved pain. But they also analysed another type of thinking: the presence of so-called control beliefs, such as 'I have learned to control my pain' or 'There are many times when I can influence the amount of pain I feel.' There was an increase in control beliefs in both groups getting therapies that targeted these beliefs — cognitive therapy and hypnotic cognitive therapy — a change that, the team concluded, was part of the reason their pain improved. Jensen thinks all those shifts in thinking represent a broader, wholesale pivot in how the brain is processing information.

Take, again, the person with chronic lower-back pain who's had serious pathology ruled out. 'Pain is not a signal of damage. It is a signal of the threat of damage. When there is no longer a true threat, it is best to turn that signal off. How do you get the brain to do that? You create new associations. When a person is hurting, the pain matrix is on. The brain has concluded there is damage happening or about to happen. It's dangerous, it's horrible, it's frightening. You want the person to stop associating the sensations they get from the back with danger. Remember, those sensations are not pain yet, they're still just sensations. The goal is to help the brain conclude, "You know what? That's normal, and that's okay." The goal is to activate the safety matrix.' Jensen wants, in other words, to turn on the constellation of activity in the brain that gives rise to a feeling of safety. Hypnosis, by coaxing the brain into slow-wave mode, and priming its neural circuits to accept a new recording, appears to do precisely that.

Yet the question remains: why can't the benefits of hypnosis be explained as a placebo effect? A story offers something of an answer. A few years back, when Jensen was still practising clinical hypnotherapy, a man came in. He was a straight-down-the-line truck driver and was, to put it mildly, reluctant. 'He said, "You know, my wife told me I have to come to see you and learn hypnosis. It's stupid, I can't do it,"' Jensen recalls. How did he do? 'I gotta tell you, he was amazed and I was amazed.' The man did spectacularly well. Then, not long after, a young woman came in to the practice. She gave off something of a hippy vibe — colourful neck scarves, exotic earrings, and an attitude to hypnosis that was all open arms and high hopes. 'I go, "Okay, this is going to be easy."' How did she do? 'No response.'

You can, says Jensen, predict who will respond to hypnosis, but not by stereotyping. It is very likely the truck driver would score higher on the Stanford scale and was probably flush with alpha or theta waves. Hypnotic susceptibility has a genetic component, too. But the story, says Jensen, also illustrates one way that hypnosis differs from the placebo effect. What you expect to happen is key to the placebo effect — believe a drug is effective and it is more likely to be. But what you expect to happen in hypnosis, while a factor, bears less relation to whether it will work for you. That's because, even though suggestions are part of both, in hypnosis they have to stick. Whether those suggestions do take root, and leverage neuroplasticity to switch the brain's response from danger to safety in people with persistent pain, may ultimately be a capability that you either have or do not.

Which is not as depressing as it sounds, because a good proportion of us could be ruled in. On one estimate, around 10–15 per cent of people are highly susceptible, and up to 70 per cent susceptible to a medium degree. These are figures that spread a little hope, and they underscore a message that Steve Olson has asked me to pass on.

'Just don't give up. If drugs aren't working, look for another way to help you. You gotta do it on your own — you don't necessarily have people out there doing it for you. If I can get one person to try hypnosis for their pain and it helps them, everything I've gone through has been well worthwhile.'

Chapter 7

The Lens of Mindset

emotion, memory, and the neurological reality of chronic pain

Head west on the US 50 from Salida, Colorado, through a valley bounded by gently cresting foothills, studded with saltbush and harbouring stands of switchgrass and prairie clover, the Tomichi Creek doing arabesques alongside as it funnels the icy melt of the Rockies past Tenderfoot Mountain to the Big Mesa and beyond, and you'll eventually come to a small town called Gunnison. Just up the hill at the back of town, standing out stark, square, and irrigated green against the Martian dun of the threadbare hills behind, is a football field. This is the Mountaineer Bowl, dynamited out of the hillside in 1948 to become the highest collegiate football ground in the world. The stadium is home to the Western Colorado University Mountaineers, a fact attested to by the team's logo — the bearded, racoon-skin cap–clad head of the archetypal frontiersman of the American West —

emblazoned big across the 50-yard line. It is not, however, just a picture. Wrapped up in that stylised image is a confluence, two properties shared between the game of gridiron and the harsh beauty of the shrub-steppes and canyons of Western Colorado: they are both breathtaking and unforgiving. And in the sweat-locker world of football, there is one position that forgives less than any other.

Back in 1988, a young man had donned the red jersey, silver helmet, and pants of the Mountaineers to play linebacker for the team. Greg Whisler was 19, had dark hair, an infectious smile, gleaming white teeth and peaches-and-cream skin, and the solid frame befitting someone who could bench press 300 pounds. He was smart, too, a maths major with a computer-science minor and an eye to becoming a credentialed high-school teacher. Yet neither wit nor weight could really prepare Whisler for what the role of linebacker would throw at him. 'I was essentially a battering ram. Back in the day, American football was a lot more run-oriented than it is today — you know, smashing heads, five yards, lower your head, get low, get low, just hit. Smack.' A given of the defensive position is that your body is up for punishment, and it was while playing high-school football that Whisler first noticed lower-back pain. At college level, the impacts were proportionately greater, and his pain went up, too.

'I would take way too much Tylenol back in the day,' says Whisler, referencing the stock pain-reliever paracetamol, whose standard dose is 500 milligrams. 'I was taking two to three thousand milligrams of Tylenol before practice. I'd have the three-hour practice, and then I would take two to three to four thousand milligrams of Tylenol after practice. I would wake up and take two to three thousand milligrams of Tylenol in the morning.' It soon became evident that Tylenol was no longer fit for purpose.

Whisler was offered stronger, narcotic painkillers, which raised a big red flag for the young footballer. 'It got to the point where I said, "I'm pretty much done. I'm not going to be a druggy." I didn't want to medicate my way through it. Chiropractic didn't work, yoga didn't work, stretching didn't work. I had multiple teammates with similar issues who had surgery. Seven surgeries were the most any of my teammates ever had, and not one of them ever got their back pain healed. I thought, "Well, surgery is not going to work, and I refuse to take the drugs. I think back pain is just a part of my life, and I'm going to have to learn to live with it."'

Now Whisler put his head down to push through the scrimmage of everyday life. He got his degree and married his high-school sweetheart, became a school teacher and then a middle-school assistant principle. He got into real estate, doing refinancing appraisals before, in the late 1990s, becoming a broker. He pursued his passion, coaching high-school football. Then he became a family man — the couple had two boys. All the while, he was dogged by pain. 'Sometimes it would knock me to ground, I mean literally, I'd fall flat on my face. It was a massive, sharp shooting pain right at my hip line, right where your belt goes across your back at the bottom of my spine. It would just send me through the roof or to the ground but always, always, always, 100 per cent of the time, it was a dull ache, a throbbing, all-encompassing tightness and tension that never went away.'

Determined to avoid the false panaceas of drugs and alcohol, Whisler turned, for solace, to food. He piled on the pounds. He stopped running and lifting weights. His activity levels plummeted, and he no longer even went for walks. The pain and immobility had a multiplier effect. 'I was to the point where, if I dropped something on floor, I thought, "How in the world am I going to pick that thing up? It is so far down there."' Years passed, and the slow rot of pain gnawed into Whisler's core, eating away at the framework that underpinned his resilience and identity. If the world looked back at Greg Whisler, it could be forgiven for seeing someone else. 'I was always of the mentality that I could toughen up and I would just deal with it. But what I didn't realise was that the pain was so pervasive, so strong, so debilitating, that it changed my personality. I became a person who was short-tempered, tired, couldn't sleep well. I became a person that wasn't very pleasant to be around.'

The family was living on a small acreage outside of Golden, Colorado, kept company by deer, rabbits, and the occasional bobcat in the open space out back, all of it overlooked by the dramatic, flat-top peak of North Table Mountain to the west. But if it seemed bucolic outside, indoors life was fraying. Whisler had become angry, aggressive, and he was lashing out. There were big altercations with his younger son over small things. His relationship with his wife began to fracture. One day in January 2014, they had an argument about the hours he spent working and coaching football at the expense of time at home. In the wash-up, Whisler, with a heavy heart, decided to quit coaching, his passion, but he was left bitter and aggrieved.

On an icy night, two weeks later, he fell asleep on the couch in the living room, and, when he woke up in the middle of the night, he couldn't move. 'I thought I just went to sleep wrong. I could not get off the couch.' After a struggle, he got to the floor. 'I literally crawled. It took me half an hour to crawl 30 feet into the kitchen.' His wife drove him to the local emergency room, where, at four o'clock in the morning, he was dosed on painkillers and given an MRI scan of his back. At 11 o'clock that same morning, Whisler got a visit from the surgeons. 'They're like, "Greg, we are fusing your back. You are jacked out." I said, "Absolutely not. I'm not having surgery."' Whisler checked himself out of the hospital, but the writing seemed well and truly on the wall. 'I was convinced I was going to be in pain for the rest of my life.'

Three years passed and, quite by chance, Whisler's wife got an email about a back-pain study being run at the University of Colorado. She forwarded it to Whisler. He applied, was accepted, and soon found himself making the 20-minute trip to the university campus at Boulder. It is an idyll of Tuscan-themed buildings in sandstone with limestone trim, awash with Roman arches and capped with a sea of terracotta, mission-style roof tiles, all set amid clusters of Douglas firs and ponderosa pines. There, Whistler was met by a researcher in the Cognitive and Affective Neuroscience Lab who ran him through a battery of tests. He had to lie on a small rectangular cushion that was inflated to raise his back. It hurt. Then he put his left thumb in a white plastic cylinder with something inside that pinched his thumbnail. It hurt, too. Afterwards, the researcher invited him to come back in a week's time for a chat with a man named Alan Gordon.

'I didn't know Alan from Adam. So Alan is sitting there chatting with me, we're chatting for about an hour, and then I had to come back in a few days. So I drive back up there and talked to Alan again, and I'm thinking in my head, "This is a really long onboarding process," and I didn't realise until two-thirds of the way through that second meeting that, you know, I'm in therapy, I'm talking to a shrink! I never really believed in therapy.' Whisler was scheduled to have eight sessions over four weeks with Gordon, a psychotherapist and expert in chronic pain who was an investigator on the study. About halfway through, however, something unexpected happened.

'Alan was running late, and I'm like, "I'm not just going to sit here in my truck." I'd checked my email, I'd answered my phone calls, so I got up and I walked around the parking lot, and I started walking down the block. Alan would send me a text saying, "I'm ten minutes away," "I'm five minutes away." I just kept walking, and I'm like, "My goodness, I've been walking for 20 minutes. I haven't walked for 20 minutes in 15 years, without pain." I just was laughing, I'm like, "Wow, this is amazing."' It was a revelation, and Whisler was stunned by the improvement, but still, he wondered, could it all be temporary? 'Later that week, I was washing my truck. I was at the car wash, and, all of a sudden, I didn't even realise it, I was down in a squatting position, squat-crawling sideways, so to speak, washing the underside of my truck. I got from the rear wheel to the front wheel, and I actually broke down crying in the middle of this car wash because I hadn't done that, or moved like that, without thinking about it, or hurting, in years. My back pain disappeared. I did not receive any pills. I did not receive any injections. All I did was sit in a small room and talk with Alan. That was my treatment.'

Greg Whisler had put up with grinding back pain, day in day out, for 30 years, and now it was gone. What happened in that room?

Vania Apkarian's life was shaped by a very different kind of pain. The Armenian genocide of 1915 displaced millions of Christians from what is now Eastern Turkey. Apkarian's ancestors were among them, and, like so many in that enforced diaspora, they found shelter in Syria, settling in Aleppo, where Apkarian was born in 1953. Yet this was not the last of the chaos and upheaval that would beset the family. The Syrian government, prone to military coups and presiding over a nation plagued by riots and general instability, would eventually lead the country to defeat in the Six-Day War with Israel in 1967. So the Apkarian family fled again, this time to Lebanon, where they made their home in Beirut. But Lebanon's own sectarian unrest was brewing the civil war that would later rive the country. So it was that, in 1971, a gangly 18-year-old found himself on the doorstep of a new world, the United States, facing a big question. What would he do now?

'I come from cultures where science was only done by gods. Of course we studied Einstein and Newton, but we thought of them as literally gods. I had no idea that I could even contemplate doing science.' All that was about to change. Apkarian, a brilliant scholar, enrolled to study maths and electrical engineering at the University of Southern California. He had not, however, found his passion. 'By the time I graduated from my degree, I decided I was not going to spend the rest of my life building circuits, and, by serendipity, I discovered that there is electricity in the brain and that my knowledge could be used to learn about the brain.' Apkarian switched tack, looking for a graduate school where he could pursue his new-found interest in neuroscience. But there was an obstacle. To be eligible, he would need to change his status from foreign student to resident, a transition weighed down with reams of red tape. 'My paperwork was messed up,' he remembers. 'The officer said, "No. This paperwork is all wrong. You are illegal now." Within a week, the FBI was searching for me to ship me back to Lebanon.' The spectre of displacement flitted into view again.

Just then, a research position opened at Upstate Medical University, part of the State University of New York system. The university, it turned out, would be able to smooth over the crinkles in Apkarian's flawed paperwork. The FBI was duly placated, and Apkarian relocated to Syracuse, New York. The brain was always his big focus, and the available position, researching pain, a mere expedient, but something about the science of pain clicked with the man who had been ripped from place to place. 'I come from many different lines of destruction on this planet, and, to me, science remains an opportunity to be optimistic about the planet, about humans, about rationality. It gives me hope that the work that we do will decrease the amount of suffering.'

That work would eventually be defined by a single question. Why, of the billions of instances where a person gets an acutely painful injury, do only a fraction go on to develop chronic pain? What set those people down the path of long-term pain? A bit over a decade ago, Apkarian set the ball rolling on a landmark study, ingeniously designed to answer that question. Apkarian was by then Professor of Physiology at the Feinberg School of Medicine, at Northwestern University, Chicago. His research team was scouting for Chicago city locals with back pain, but there was an important rider. The pain had to be of recent onset — in the last four to 16 weeks — with no other episodes of back pain in the preceding year. The plan was this: the team would track those people for a full year, get them to rate their back pain while having a series of brain scans, and see what differences there were between people who got better and those whose pain persisted. Which might sound straightforward, but there was one major hurdle.

'If I ask the patient right now, "How much pain do you have?" let's say the patient says, "I have five-out-of-ten pain." If I ask that same question one minute later, it is essentially a random number generator. It can be anything,' says Apkarian. 'Pain itself is not a stable concept. It is continuously shifting in its magnitude.' This was a big problem for Apkarian. He wanted to measure where the brain was active when people reported back pain. If he didn't know when they were in pain, or how bad it was, the measurement would be impossible. Which is where the maths and engineering nous that Apkarian honed as an undergraduate became very useful. He had devised a special gadget that used position sensors — electrodes attached to the person's thumb and index finger — that let the person report pain by simply moving those digits. Thumb and index finger touching in a pincer grip meant zero pain on a 100-point scale, while stretching the digits fully apart rated the full 100 points. And, of course, there were all the gradations in between. The gizmo — allowing a continuous report of pain levels, which could be seamlessly time-matched with brain-scan activity — was enlisted for service.

At the end of the recruitment drive, 94 people made the grade. The volunteers, a little over half of whom were women, had experienced an average pain intensity of six out of ten for a mean of nine weeks. One by one, they arrived at the Feinberg School, which is fronted by the Montgomery Ward Memorial Building, a towering neogothic structure of Indiana limestone over concrete and steel, built in 1927 and known as the world's first academic skyscraper. In Apkarian's hi-tech lab, each person had their initial brain scan, all the while shifting their thumb and index finger back and forth, continuously rating their back pain with Apkarian's gizmo.

'When they rate their pain, the brain regions that are involved in that rating are the same brain regions that would light up if we had pinched their skin. All the brain regions that we know are activated when there is an acute painful stimulus were identified in these subjects. So we say, "Okay, this is like acute pain."' Not surprisingly, the areas lighting up centred on the sensory cortex, widely accepted as the location where sensations, including pain, are felt. Who had those changes? Unerringly, it was everybody, all 94 of them. 'They all have acute pain, and all of it looks the same.'

But Apkarian was about to roll the dice. Remember, these were all people in the early stages of back pain. Apkarian had no idea how many, if any at all, would go on to have persistent pain. If no one got chronic pain, the study would, in many respects, be a waste of time. As the months passed, the fall of the dice became apparent. 'We were lucky in the sense that approximately 50 per cent of these patients over the span of the year continued to have exactly the same pain, while the other 50 per cent, their pain went down,' says Apkarian. 'So we had this clear dichotomy of patients.'

During that time, Apkarian's subjects diligently tramped back to the Feinberg for pain ratings and brain scans at three-, six-, and 12-month intervals. On those visits, the scans of people whose pain went away were more or less regulation. 'The intensity of the pain is less, so over time the brain activity just disappears, and that makes perfect sense,' says Apkarian. 'In these subjects, the brain activity goes down to zero over that one-year time period.' What about the people whose pain didn't go away? As you might expect, their brain activity continued to light up the scans. But something else did change. Dramatically. Their pain was no longer being registered in the sensory part of the brain. It had shifted to a completely different set of regions:

the amygdala, the hippocampus, the medial prefrontal cortex, and the nucleus accumbens. Weirdly, these were regions not typically linked to the perception of physical sensations. What was going on?

To begin to understand, we need to head back in time, to 1939, and westwards, to the University of Minnesota, where the celebrated psychologist B.F. Skinner had just invented a device that, for its future inhabitants, would be none too pleasant. It was a series of 24 cylindrical boxes, each equipped with a metal grille floor, a lever poking from the wall about halfway up, and a short brass tube at the back. The whole thing looked like a cross between a telephone switchboard and a roosting coop. It was, in fact, a version of the famous Skinner box, painstakingly designed to study animal behaviour. Together with a young psychologist named William Kaye Estes, Skinner began a series of experiments with a group of male albino rats. First, the pair trained the rats to press the lever by rewarding them with food pellets, specially made by pharmaceutical company Parke-Davis from a choice selection of grains, which came out of the little brass tube. Rats duly trained, the researchers began to sound a tone, twice every hour, via a telephone receiver in the rear wall of each cylindrical prison. The tone went for three minutes, at the end of which the rats were startled by a shock, delivered to their paws via the metal floor, which was electrified. Over the next few days, the pair lengthened the tone to five minutes, and watched the animals' behaviour with hawklike intensity. Soon, they began to notice something peculiar. The rats were pressing the food lever less whenever the tone sounded — there was a

75 per cent reduction early on, and then, around the end of the third day, the pair reported that 'responding practically ceases during the presentation of the tone'.

They rats were, quite simply, afraid. They had learned, been conditioned, to fear the tone coming from that telephone receiver. 'If we go to the conditioned-fear learning literature, all of it is based on a shock, on a painful stimulus,' says Apkarian. 'So it's the memory trace of pain, that is what they are studying. They call it fear, or they call it anxiety, but all of those mechanisms are definitely relevant to the long-term traces that a painful stimulus is going to generate in the brain.' An innocuous sound had become indelibly linked in memory to a painful shock and could now, all by itself, produce the classic animal fear response: freezing.

Which is why one of Apkarian's findings is so eye-opening. The amygdala, an almond-shaped knob of tissue deep in the brain's temporal lobe, consistently lit up in his persistent-pain group. It is the very core of the brain's fear system. The amygdala can, at a moment's notice, flood a person with anxiety and bring on the sweats and palpitations of panic. The amygdala is also key to the acquisition of fear memories, especially those linked to cues such as the tone that Skinner played to his rats. Or, indeed, the cues that were present when a person first hurt their back. For example, they might have been bending forward and lifting. The pain of the injury can become linked to the cue that came before: bending forward. If the person bends over again in future, or perhaps just anticipates bending over, that movement can trigger fear, all by itself.

The hippocampus, meanwhile, is perhaps the key player in the formation of memories. Apkarian's research has shown that, in persistent pain, the hippocampus undergoes subtle changes in shape that go hand in hand with an enhanced memory for pain. Which can make things uncomfortable in ways you mightn't expect. 'If I walk into a room and it just happens to be that my back pain is high at that moment, given my back pain fluctuates up and down all the time, next time I walk into that room my pain will be even more exaggerated, just by the existence of that memory trace within my hippocampus, which will now act on the nociceptive signal coming in, and amplify it,' he says.

But the role of the hippocampus, like so much about pain, is complicated. The hippocampus doesn't just harbour pain memories, but, like the amygdala, also holds our fear memories. And each part of the brain has its own patch. If the amygdala stores fear memory for cues, such as a tone or a movement, the hippocampus stores fear memory for places. It is the hippocampus that's the guilty party when fear becomes linked to locations, a worksite for example, where a painful injury happened. All of which suggests that buried deep in the mechanism of chronic pain is a fear memory that can be activated by something as simple as a movement or a place. The conclusion is as compelling as it is disconcerting: fear causes pain.

'Limbic, emotional brain areas, primarily prefrontal cortex and areas like the hippocampus, amygdala, accumbens, are all the areas that are becoming more active. The people getting chronic pain are activating more and more emotionally laden brain areas,' says Apkarian. And there is something unique to the kinds of emotions being generated by people with chronic pain: they are all negative. They include fear and its close cousins, depression and anxiety. Negative emotions have always connoted some form of threat. Imagine how you'd feel if you were about to run out of food, be evicted from your house, or be left by your partner. Pretty lousy, I'd guess. Those downer emotions evolved to make you view each issue as a threat, and then act to prevent bad things happening — all of it motivated by seeking relief from negative feelings, and gaining the reward of positive feelings should you be successful. It follows that, if you tag a sensation from the body with negative feelings, your brain will conclude that the part is under threat. It is in danger. And so your brain errs on the side of feeling the sensation as pain, prompting you to protect the part by, for instance, not moving it. Better safe than sorry. 'Pain is an emotional state, right, it is a negative affective state, we cannot even dissociate those from each other,' is how Apkarian sums it up.

The wash-up, as Apkarian wrote in a recent paper, is that chronic pain is not 'the temporal extension of acute pain'. The perception in each case is identical — it hurts — but the brain mechanisms couldn't be any more different. Acute pain resides in the sensory cortex. Chronic pain dwells elsewhere, in the emotional parts of the brain. They have pain in common — beyond that, the chronic version is a different beast altogether.

What sets apart the 50 per cent of participants, from Apkarian's broader pool of subjects, who got chronic pain? 'We literally looked at the brains of these subjects at time zero and said, "Is there something in the brain that tells us who, one year later, is going to continue to have pain, versus to recover from pain?" That was amazingly successful. We were somewhere between 85 to 100 per cent correct in predicting who, one year later, is going to have pain versus not.' The key lies with the remaining two brain areas whose activity was so prominent in the chronic-pain group: the medial prefrontal cortex and the nucleus accumbens. Those areas have a big role in something you might never have guessed had any link to pain. They are part of the reward system, which can motivate us to eat tasty food or try hard at work, but which also, crucially, keeps some people addicted to drugs, alcohol, and gambling. Apkarian's team found that people with stronger connections — enhanced activity between these two areas — were the ones whose pain would 'chronify'. It sounds absurd, but something in the experience of pain was delivering a reward in the same way as an addiction.

It turns out that if you've got persistent pain, *avoiding* things that make pain worse becomes a reward in and of itself, flooding the brain with the happy chemical, dopamine. To get that reward, people will go to extraordinary lengths to avoid certain movements, and can become obsessively vigilant for any sign the pain is coming back or getting worse. It's behaviour that looks and smells a lot like an addiction. I'm reminded of Jane Trinca's patients, whose strategies to position their body just so with an arrangement of pillows, or only sleep in a certain chair, or indeed stop moving altogether, seemed to verge on the irresistible.

What are you left with? A very big carrot and stick. Don't

move and you get the reward of 'no pain', something termed 'positive reinforcement' in the psychology literature. You also avoid a punishment — pain — which is called 'negative reinforcement'. Both are powerful motivations to stop moving. But, and here's the kicker, people who are wired to get the most reward from avoiding pain are the ones most likely to get persistent pain — because they are also the ones wired to see pain itself as a huge negative. If you see pain as something really bad, its presence, or the risk of it, will naturally produce all those negative emotions, the ones that put the Picasso Blue Period filter on the signals coming from your back. 'You have positive and negative reinforcement-based emotional valuation of your environment, and that circuitry predicts the long-term future in these patients,' says Apkarian. 'In a way, their decision-making machinery is their vulnerability for developing chronic pain. That circuitry is deciding the extent of amplifying or de-amplifying the nociceptive signal that comes to the brain. In chronic back-pain patients, we think this central amplification of the nociceptive signal is the primary controller.'

Apkarian's work offers a whole new way to look at pain. But it is also scary. I remember reading through some of his papers before emailing him for an interview. It was depressing, anxiety-provoking. If chronic pain is a persistent memory trace, rooted in fear, can it ever be extinguished? If chronic pain is like an addiction, if you can predict who's going to get it, does that make it a sentence set in stone? Is it simply a destiny that, for each of us with the misfortune to keep hurting after an injury, is etched in the stars? And if it's bound up in the emotions, out of reach of drugs and surgery, what can anybody do about it?

Head south from Yellowstone National Park in Wyoming, with its craggy, snow-lined peaks, plunging canyons fringed with lodgepole pines and blue spruce, swollen cataracts that belch tumbling spray into boiling rivers below, and hot springs that pock the ground like eyes whose tears fall upwards, and you come to a high valley, a geographic feature coined, by early fur trappers, as a 'hole'. This is Jackson Hole, at the south end of which is the town of Jackson, nestled between the Teton and Gros Ventre mountain ranges and wrapped around with some of the best skiing in America. This was Tor Wager's stomping ground as a boy.

Growing up in the 1970s and 1980s, he was steeped in the rarefied air of the high country, but also in the esoteric teachings that came with having a mother who adhered to a very particular faith. 'My mother, when I grew up — and I also went to this church as a kid — was a Christian Scientist, and Christian Scientists are famous for believing in the power of prayer and the mind over the body, especially to heal. And so I grew up surrounded by a lot of people who had a very strong belief in the healing power of thought,' says Wager, who is director of the Cognitive and Affective Neuroscience Lab at the University of Colorado. 'I never bought into the mystical part, but I've always had an interest in how deep this could go. What can your thoughts and beliefs affect?'

After finishing school, Wager began probing paths that

might deliver answers. He enrolled at Principia, a Christian Scientist college in Illinois, where he started a philosophy and physics double major before switching to a Bachelor of Music Composition. He was also honing a mental practice with yoga, a discipline that brought with it a pull to the East. After graduating in 1996, Wager travelled to Asia and found himself hiking the mountain paths between the enormous flint-like peaks of the Annapurna massif in Central Nepal. There is something about the yawning vacuum of an impossible expanse that focuses the mind, but something else happened on the trip that made things crystal clear for Wager. 'I had eaten something bad. I got off the beaten path, eaten some bad dal bhat, got sick, and spent about a day doing nothing but just sitting with a fever at this little picnic table, at this little hostel, with nothing to really do,' Wager remembers. 'But at the same time, I was having this sort of crisis of meaning, essentially, what am I going to do? I thought, "What's the coolest thing to study?" Then I thought, "This is the brain."'

When it comes to the brain, and the power of thought to heal, two subjects naturally gravitated into Wager's orbit of attention. Placebos and psychotherapy both treat pain, and they both do it by getting a person to think differently. They also share the same problem — there is a great deal of uncertainty about how each works, mechanisms that, because of their spongy, elusive nature, Wager refers to as 'squishy things'. 'It's really hard to quantify them,' says Wager. 'You don't know what the dose is. You don't know what the active ingredients really are. And when you don't know how a treatment works, there is something else you do need. It's all the more important to have really good targets. Are they affecting hard outcomes, so to speak, are they affecting biological, physiological outcomes?'

What kinds of outcomes might those be? Sure, you could simply ask a person, 'How's your pain?' But, as Vania Apkarian explained, pain reports are notoriously fickle. Wager wanted to know this: was there a diagnostic test for pain that didn't depend on a person's say-so? Was there an objective marker of pain in the brain?

In 2010, Wager was an associate professor, installed in his own Tuscan-themed building at the University of Colorado, and he assembled a research team to find out. They recruited 20 healthy people and strapped thermode devices, programmed to deliver heat pain at a range of intensities, to each volunteer's left forearm. The researchers asked each person to report when they rated the pain at a one, three, five, and seven on a nine-point scale. Then the team switched tack. They calibrated the thermode to give each person those exact same pain intensities all over again. This time, however, the researchers were scanning the volunteers' brains with an fMRI machine. They collected image after image as each person experienced one, three, five, and seven pain. Which is when they cranked the tech up a notch. The team now deployed a machine-learning algorithm to comb through the scan slices, each with data on thousands of brain locations, hunting for a pattern of activity that matched each level of pain intensity.

In an article in *The New England Journal of Medicine* in 2013, Wager's team reported that the algorithm, in an extraordinary result, had delivered. When they applied it to a new group of volunteers, they could predict, with 93 per cent accuracy, whether those folk were feeling a one out of nine on the scale — 'non-painful warmth' — or a three, five, or seven out of nine — all levels of painful heat — just by analysing the subjects' brain scans. Wager's team dubbed their brain template the Neurological Pain Signature. The brain areas getting active were those you'd expect, classic areas involved in nociception and the response to acute pain, including the sensory cortex and an area called the insula.

Wager had taken a long stride towards an objective marker of pain. But there was still another question hanging about the wings: what happens in the brain when you give a placebo analgesic, when you convince someone that a fake drug or treatment will help their pain? In 2015, Wager teamed up with his former graduate student Lauren Atlas, who was now a pain expert at the National Institutes of Health. Together, they trawled through hundreds of studies that had gathered brain-imaging data on people given placebo painkillers. The result was a striking set of images of the brain, studded with blue dots. The dots were areas where brain activity had gone down after the placebo — pain regions, including the sensory cortex and the insula. Placebos looked to be hitting pain precisely where you'd expect. But the pair's brain artwork also featured another bunch of dots, coloured red. Here, brain activity had, curiously, gone up after people got a placebo.

One standout red area was the ventromedial prefrontal cortex, or vmPFC. Wager knew just how important this finding was. Studies in animals have found that stimulating the vmPFC can shut the gate on incoming pain signals to the brain. But if the area sounds familiar, that's because it is — it is almost exactly the same region that was active in Apkarian's patients who went on to have persistent pain. Why would an overactive medial prefrontal cortex be linked to more pain in Apkarian's patients, but less pain in people who got a placebo? Wager put together a thermode study that would clear things up.

This time, his team asked people, as the thermode heated up on their forearms, to use their imaginations to make the pain seem worse, prompting them to, 'visualise your skin sizzling, melting, and bubbling as a result of the intense heat'. Then they coaxed the volunteers to make the pain seem better by calling it 'pleasantly warm, like a blanket on a cold day'. Those thinking strategies shifted pain ratings in the expected directions. But as it all played out, each person was having their brain scanned, and it was those results that were most intriguing. 'When we look at training people to appraise pain as either terrible, burning, awful, or okay, tolerable, "warm blanket on a cold day", and so on, it's the ventromedial prefrontal circuit, not classic nociceptive circuits, that mediate those changes in thinking, on pain ratings,' says Wager.

Wager's team had found that greater activity in the vmPFC could turn pain both on and off, depending on how each person was interpreting the heat on their skin. The vmPFC, it turns out, is a key player in a hive of brain circuitry called the default-mode network, which does something remarkable. The default-mode network is critical when we discern value, when we work out what a situation means for us and our wellbeing.

For people with chronic pain, that includes deciding whether a sensation is merely harmless and uncomfortable, or dangerous and intolerable. 'The vmPFC is not simply "good" or "bad" for pain,' says Wager. 'We think it provides the meaning context in which pain is interpreted, the "lens" of mindset. In a healthy person, it's "pain off". In a person with [persistent pain], the activity that generates this context switches to being "pro-pain".'

Wager was being led to an inexorable conclusion: the kinds of reappraisals people make when they get a placebo, or when they get psychotherapy to reinterpret painful sensations, were tamping down pain levels by recruiting the vmPFC. If that was right, it could, Wager realised, address something that had been bugging him for a while. 'We've had some success tracking brain activity related to how much pain are you in right now,' says Wager. 'But that is not chronic pain, because chronic pain is a condition of a person, it is not an immediate experience, right, and pain isn't the only thing involved.'

Wager had a new Holy Grail. Could there be a brain biomarker, a neurological signature, for people in chronic pain? If there was, did it, as Apkarian's work hinted at so strongly, involve the vmPFC? Most important of all, could you recruit the vmPFC and switch it to 'pain off' with a placebo? Or, just maybe, with a specially tailored form of psychotherapy?

In the northern spring of 2017, the redbud trees at the University of Colorado were blooming fiery pink, and Wager's team was in full recruiting swing for a back-pain study. They had placed ads on Facebook and Craigslist, in the local rag the *Daily Camera*, in leaflets at local pain-management clinics, and it appeared in Google searches for chronic pain. The researchers were looking for locals aged 21–70, in the Boulder–Denver region, who'd had back pain for most of the previous six months. Winnowing complete, 151 people, average age hovering around 40, were signed up. Among their ranks was a middle-aged man called Greg Whisler.

The researchers set about dividing Whisler and his fellow volunteers into three groups. Fifty would stay on their usual treatment, which, says Wager, was generally anti-inflammatory drugs and opioid painkillers. Fifty-one would get a placebo. Those people were treated to a couple of videos with a key message: placebos can powerfully reduce pain, due to 'engagement of the body's natural healing responses'. Placebos work, they were informed, even when people know they're getting a placebo. That's right — the volunteers knew they would get a fake treatment. It's called 'open-label' placebo, and a number of studies show it can be as effective as a 'deceptive' placebo, where the person thinks they are getting the real drug. People in this placebo group got an injection of one millilitre of salt water just under the skin of their back, right at the point it hurt most. Whisler, of course, was in the third group, 'chatting' with the psychotherapist Alan Gordon. The chats were, as Whisler eventually realised, a uniquely formulated psychotherapy, which the researchers called Pain Reprocessing Therapy.

The team followed people for a full year before crunching the numbers. Their primary question was this: how many people were pain-free or nearly pain free — that is, registering a zero or one on a ten-point scale of back pain — after one month? Perhaps unsurprisingly, of the people who'd got 'usual care', only five — 10 per cent of the group — hit that benchmark. The people who'd got the placebo did a little better, with ten people in that group, 20 per cent, registering no or nearly no pain after a month. But when Wager's team rung up the numbers on the group who'd had Pain Reprocessing Therapy, they found something extraordinary. 'People beforehand had to be a four-out-of-ten pain to get into the study, and they'd been in pain an average of ten years. After the study, 66 per cent were pain-free or nearly pain-free, and the effect was diminished very little after a year. So people really stayed better,' says Wager. 'The news of this paper is that there was really a massive effect of the psychotherapy versus the placebo or usual care.'

It was a stunning result. Whisler was by no means an isolated case — in his psychotherapy cohort, fully two-thirds of participants were making dramatic recoveries. But how was Pain Reprocessing Therapy working? The answer lay in one of the batteries of tests they had taken, something called the Tampa Scale of Kinesiophobia. As the name implies, it assesses fear of movement. People who score high on the Tampa Scale agree strongly with statements like *My body is telling me I have something dangerously wrong* and *I'm afraid that I might injure myself if I exercise*. 'If you look at the items, what they're about is people's perceptions of the causes of pain and what it means,' says Wager. 'For example, "If I feel pain, that always means my body has been injured." That is really a belief about what the pain is, what it is caused by, and what it means for your future.

"If I try to overcome it, the pain will only get worse," "I will be in pain for the rest of my life". Those were the beliefs that showed the strongest change with the psychotherapy and were the predictors, the mediators, of efficacy.'

What was it that Gordon and Whisler talked about in that little room in Boulder? It turns out that their consulting room chat was focused on a list of key messages. Gordon taught Whisler about central sensitisation — that pain is mostly a useful sign of damage, but sometimes it's a false alarm, 'there is no fire', and the body isn't really injured. He supported that with evidence, such as how abnormalities on a person's back MRI are often seen in people with no pain. Isn't it interesting, Gordon would suggest, how pain gets worse with stress, an emotional trigger that has nothing at all to do with movement? He would home in on sensations from the back and encourage Whisler to reappraise them: 'Even though it's a tight, burning, tingling sensation, we know that it's safe. We've gathered a lot of evidence. Your back is … healthy. Your brain is simply misinterpreting the signals coming from your body as if they're dangerous.' Gordon would use techniques from mindfulness, suggesting that back sensations need not be changed, or gotten rid of, but merely observed. Like any good psychotherapist, he explored other issues in Whisler's life that might heighten his sense of stress, fear or threat. But something else important happened during therapy. Outside of the little room.

'It's followed up with exposure, which means — do the thing that hurts. If it hurts, it's totally safe, go ahead and move,' says Wager. 'So there's a real commitment there, it's very gutsy. It's the suggestion and belief, and then it's putting that into practice through movement and exposure — self-exposure to pain and a return to activities. So you're almost proving to yourself that this proposition is true, that it's just pain, right, it's safe.' In short, it was a neat, expertly delivered distillation of some of the main messages from modern pain science and psychotherapies such as cognitive behaviour therapy.

Whisler described his recovery as 'miraculous'. Wager, however, was banking on something more earthly. The team gave everyone brain scans at the start of the study, while participants had pain induced with the inflatable cushion that stretched their back upwards, and when they just reported spontaneous pain, minute by minute, on a visual analogue scale. The scans were repeated five weeks later, after the treatment had finished. On those second scans, two brain areas showed reduced activity. 'They are centres, the insula and the anterior cingulate, that are the most consistently engaged by painful things. In our study, we were able to measure the activity in these regions as tracking moment-by-moment back pain and then normalising with treatment. We think that's really a correlate of this immediate, nociceptive pain experience, which is reduced after treatment,' says Wager.

Wager had hard evidence of brain changes reflecting reduced pain. But what role was the vmPFC playing in all this? To explain, Wager suggests a thought experiment. 'Try to imagine that somebody was breaking into your room right now with a gun, and your life was in danger. Could you make heart rate go up? Yeah, maybe you might be able to talk yourself into being a little nervous, right?' he says. 'But if you truly believe that, if events conspire to really make you truly believe that, it would be a completely different world.' Wager's example is all about reappraisal and a very big question that comes with it: how does a person come to genuinely believe in their new world view? He thinks we need a special ingredient to make it real, something supplied by the vmPFC and its hive of brain circuits. 'All the arrows are pointing to the idea that the default-mode network is a centre for constructing meaning, a sense of self and context, and it integrates various bits of context from memories, from the world around you, into a conception of what are the causal events that I'm supposed to be inferring from my experiences, and how does that bear on me and my wellbeing. And when something does bear on you, and really engages the system in the right way, then that becomes emotional to you.'

Just as pain is made menacingly real by the dark emotions that accompany it, Wager thinks that pain reappraisals are also made real by emotions — positive ones. It is an 'affective appraisal' where the proposition 'my back is safe' isn't merely intellectual, but is fully embraced by the emotions with which we imbue the world with value. The new thinking now becomes authentic, real. The vmPFC, with its byzantine connections that snake into the brain's pain-sensing systems, exerts an emotional sway that can shut down pain. So did Wager find a signature of chronic pain? 'We looked for changes in the vmPFC, and we didn't find them,' he says, with a hint of disappointment. Why not? 'There is no guarantee we are going to see those signals in the scanner. When the meaning change is happening is not when they're being scanned — we need to know the precise moment to capture that.' For Greg Whisler, it seems fair to conclude, that 'meaning change' happened in the days before his 20-minute walk around the block and his squat-crawling epiphany at the local car wash.

Uncertainties aside, Wager's construal shares much with that of Apkarian. Fear is the key. Fear of pain, and fear of movement, filter innocuous sensations through a lens of threat, transforming them into the danger signals of pain. But fear conditioning can be reversed. New beliefs — grounded in evidence and reinforced by the experience of movement and the positive feelings that come with new understanding — can extinguish pain.

Not long ago, Wager gave a talk to a group of anaesthetists. He asked them to consider two hypothetical patients. One has recent onset of pain in the knee. Imagine, he prompted the audience, you have a new painkiller, something in development called a 'Nav 1.7 channel blocker', which could block signals from the nociceptors in the knee. This might work, Wager told them, because 'this person has peripheral pathology.' The pain is caused by damage in the knee, so a treatment that targets the knee makes perfect sense. Then Wager invited the doctors to consider a second patient, with knee pain that has become chronic. The pain has shifted to the emotional brain regions, the prefrontal cortex and accumbens. What does that mean for treatment? Wager was categorical: 'There's nothing you can do to the knee that will ever make them better.' That person, he concluded, needs a pain-focused psychotherapy, like CBT or Pain Reprocessing Therapy.

This is the challenge of maladaptive neuroplasticity, wrapped up and handed over in a neatly swaddled package. Persistent pain hurts. It is real. But it dwells in a different place, often in the emotional areas of the brain, and so any treatment needs to have the emotions firmly in its sights. And there are a great number of reasons why it is critical to get that target right.

'I'll be open and transparent, and as honest as I can,' Greg Whisler tells me. 'I was married to my high-school sweetheart, 16-years-old I started dating her. I had two amazing children, who are now 21 and 19 years old. They were my world, that was everything I ever wanted, and I love them dearly. But the chronic pain infected my personality. Like I said, I became a person that nobody wanted to be around, to the point that — and I'm sure there were other things — but unfortunately, a year ago, my wife left me after 26 years of marriage.

'My younger son and I have altercations, and I think some of it is just my short-temperedness, and my lashing out, mostly in pain, not really at him. I mean, the things that we argued about or fought about really could have been responded to in much nicer, less confrontational ways. But because of where I was emotionally and physically, pain-wise, it came out as anger and aggression that drove them away. So the four of us, that I used to say was my family, two of them don't speak to me anymore, and that devastates me. Ultimately, it cost me everything that was most dear to me.

'In the work that I have done with Alan, I realised I did not have to live with daily pain. I could actually live a normal, fully-active life. Pain-free. It starts to change your perspective on life, to the point where goals, dreams, desires actually become a reality, instead of the dreaded, "How am I going to make it through today because it hurts so bad?" Unfortunately for me, in my case, it was too little, too late, and now I am here picking up the pieces, so to speak, and reassembling my life in a manner to make sure that doesn't happen again.'

One of Whisler's passions was riding motorcycles. He hadn't ridden one for 15 years. Ten minutes sitting on a motorcycle and the pain was just too bad. Not long after his sessions with Gordon, he took it up again and started doing rides of several hundred miles. It was elating, and, buoyed by those shorter trips, he and a friend planned an epic 'four corners' tour of the United States, taking in the compass tips of San Ysidro, California; Blaine, Washington; Madawaska, Maine; and Key West, Florida. Covid put the brakes on the full tour, so they reined it in to just the southern states.

Part way through the trip, in the dark, early-morning stillness of a late-May morning in 2020, the pair left Jacksonville, Florida, to head west on the Interstate 10. As the sun rose, they motored through Tallahassee then Mobile, Alabama, and Baton Rouge, Lousiana, stopping every three hours or so for fuel and a leg stretch. They kept riding until sundown, then into the night, and they were in Texas, still on the Interstate 10, closing in on the tiny town of Van Horn. There, the landscape is mesmerically flat, cattle-ranch country scarred white by sand and limestone quarries, the monotony broken only by the occasional undulation of distant hills. The hypnotic effect is magnified after dark, when the clear air, unsullied by city lights or tailpipe fumes, makes the stars tantalising and bright against the black night. Just after midnight, the sky was cloudless, and, lulled by the pulsing thump of the motorcycle engine, Whisler took his eyes of the road.

'The stars were absolutely amazing, and I got distracted. I started staring at the stars on this little stretch of what I thought was straight highway, and I saw the Milky Way, and I — literally, this is making me emotional — turned my head for a split second too long.' As the Interstate 10 comes up to the Hoefs Road turn-off, it makes a very slight bend to the left, which Whisler, in a fateful moment of stargazing, failed to negotiate. He left the road at 80 miles an hour, and, as the bike hit the loose, roadside dirt, he was thrown off, crashing into the ground on his left shoulder. Whisler was stunned, but, in shock and with adrenaline surging, he made it onto all fours. 'I just started walking back to the bike, and my buddy finally caught up to me. He said I was Dead Man Walking. He tried to yell at me, and I couldn't hear.'

They airlifted Whisler to the trauma centre in Odessa, where his injuries were methodically catalogued by the ER team. He had a shattered collarbone, eight broken ribs, a punctured lung, a broken shoulder blade. 'I crushed the whole left side of my body. I had lots of surgery, broken bones, a week in ICU, and a month of bedridden,' Whisler remembers. But there was something else about that hospital stay that was notable, in its absence. 'My back never hurt once. Absolutely blew me away.'

In one of their sessions, Alan Gordon had given Whisler a

handheld clicker, like the ones they use at turnstiles to count how many people are coming in. Whisler should use it, said Gordon, to check off every time during the day he would think about his back pain. The idea was to train Whisler to notice the sensations from his back and describe them, something he often did in terms of temperature and colour. Now, lying in his hospital bed, Whisler had no clicker, but his training held good, and he analysed the stutter and crackle of sensations this new suite of injuries was radioing in. 'The rib pain tended to be black and blue, very dark blue, very spotty, very dull, achy, where the pain in the back of the ribs and the shoulder blades and even the collarbone was much more sharp and needle-like and red and orange,' he says. 'I would just think about it, describe it and feel it, not trying to change it, not trying to stop it, but just observing it.' The technique, drawn from mindfulness, moves away from engaging with unpleasant symptoms, to a stance of detached, non-judgemental watching, which can de-escalate things. 'I didn't have to deal with back pain at all, so I just applied those same techniques looking at the other sources of pain.'

The University of Colorado back-pain study, published in *JAMA Psychiatry* in September 2021, is impressive. There are always grains of salt, of course. Applicants were, for example, excluded if they had symptoms that suggested neuropathic pain, which may be less sensitive to psychotherapy. And there is something pretty special about Alan Gordon, too, as Tor Wager attests. 'What I think we don't know about Pain Reprocessing Therapy is who it's going to work for, under what circumstances, and how committed the therapist needs to be. So Alan Gordon did almost all of these treatments, and he's the kind of guy who'll say, "I'm going to make you better, we're going to do this together."' On a ten-point scale of commitment, Gordon rated an 11, and — as Luke Chang, who ran the red Vaseline/ blue Vaseline study with Wager, argued so convincingly — the allegiance effect is prominent in almost any recovery.

None of which changes the fact that, when you understand that persistent pain springs from a part of the brain that plays very little role in the perception of acute pain, you ought to hold a different view on how to treat it. A little while back, Vania Apkarian gave a talk to a group of surgeons, each of whom saw a good number of patients with persistent lower-back pain. Apkarian listened first to several of their presentations, which contained detail after detail of a litany of surgical techniques to treat back pain. 'In fact, they were all spine surgeons, and they showed one picture after the other of all these surgical procedures,' says Apkarian. 'My message is very different, in fact.' It is a message directed squarely at all those people considering surgery for back pain, whose brains are no longer registering it in the acute regions, but in the emotional areas that lit up so clearly in his, now classic, study. 'Just don't do it, kind of thing. Right?'

Conclusion

Become a Bowerbird

Head south on Springvale Road in Mulgrave, in the outer suburbs of Melbourne, past McDonald's and over Wellington Road, and on your left you'll find a clutch of streets called Glenvale Crescent, which is, intriguingly, rectangular. Take a gander around its neat geometry and, chances are, you won't notice much. It is dreary, industrial, a conglomeration of low-rise warehouses, in cream and brown brick from the '60s and '70s, big roller doors down the side and nondescript glass-panelled offices out front to welcome wholesale customers. There's a shopfitter, a purveyor of printer consumables, an office-furniture supplier, and sundry others, all deathly quiet apart from the occasional forklift truck, busily scooting pallets of goods, or, here and there, a fluoro-clad tradie emerging from a work truck. On a Sunday morning, however, all of that changes. The Glenvale rectangle, measuring near enough to one kilometre in length, transforms into a venue for something altogether more spectacular.

Between October and April each year, the Carnegie Caulfield Cycling Club cordons off the square of streets for a series of criterium bike races. Over the years, they have featured some of the world's top cyclists. Tour de France winner and Aussie cycling legend Cadel Evans raced there in his up-and-coming days, back in 2000. For the record, he came fourth. Baden Cooke, who took the sprinters' green jersey at the 2003 Tour, has also done time on the Mulgrave circuit. Simon Clarke, who won king of the mountains at the Vuelta a España, has been a regular at the Glenvale crits, too. And, on a sunny February morning in 2010, as Clarke geed himself up for the A-grade race, a less well-known bike racer rolled on the lycra and saddled up to participate in his first-ever D-grade event. That would be me.

I was 47 years old, had been doing recreational road riding for six or seven years, and had put bike racing on my bucket list. My inaugural crit would be 45 minutes plus one lap of the crescent. I was nervous but, wired with adrenaline, started off with modest hopes of staying with the bunch. There was some early drama: a crash on the turn coming up into the service lane on Springvale Road. Pretty soon, though, I'd found my rhythm and somehow managed to be ahead of the pack. This felt good for a while until I realised, a lap or so before the bell signalled the final loop of the crescent, that I had miscalculated, firing too soon. The bunch caught me, and its lead rider screamed, 'Is there anyone ahead of you?' 'No!' I yelled back, wind whistling in my ears as a thunderous whirr of bikes and riders slipped past, leaving me, spent, to roll home among the stragglers.

It was a great lesson, and addictive. I decided to train harder. In those days, I'd ride my workhorse bike, fitted with two big panniers, to the Queen Victoria Market for the weekly shopping. Sometimes I'd come back with 25 kilos in the bags.

Why not, I thought, use the time to do some sprint training? With the added resistance of a leg of lamb, some apples, maybe a pineapple or two, and a watermelon, I started doing interval sprints along Beaconsfield Parade, the palm-studded foreshore strip that leads from my home in Bayside. But not long after this grocery-fuelled training ritual began, I noticed a pain above my left kneecap. I rested and it got a bit better, so I started riding again. It got worse. I stopped riding. Then the pain started to spread, around the side of my knee and under my kneecap. Running seemed to make it worse, so I stopped running. I went swimming, but the freestyle kick made it worse, so I stopped swimming. My exercise now consisted solely of walking. This went on for a couple of months before I mentioned it to my physio friend Paolo, and he took a look at me.

After much judicious poking and prodding, Paolo concluded that I had a case of something called patellofemoral-pain syndrome. It's an overuse injury, often seen in runners and cyclists, thought to be caused by stress where the back of the kneecap, the patella, meets the front of the lower end of the thigh bone, the femur. Paolo suggested some exercises. Alas, they didn't help. The pain was never agonising, but I often had to sit with the leg straight, and always, at the back of my mind, was the thought that I might never race or even ride my bicycle again.

It was only after several months that I decided to see a sports-medicine specialist. Which was when I booked in for my first consult with Kal Fried, a decade before I would see him for the other knee. It's a while ago, but two things always stuck in my mind about that visit. 'People who sit on the couch don't get better,' Fried told me. This was the opposite of all my medical training, which said that injuries should be rested, iced, compressed, and elevated — neatly summed up in the acronym RICE — until they got better. Fried also mentioned the term 'neuroplasticity' and explained that, as my pain was now officially 'chronic', sensitisation in my spinal cord and brain would, almost certainly, play a role in perpetuating it.

Fried suggested I get a stationary trainer for my bike. I did. It was an orange-and-black wedge-shaped metal frame that sat on the floor, with a roller to rest the rear wheel of the bike on, which let you ride as much as you wanted in the garage, without going anywhere. Start low, then gradually build up, Fried advised. Ever cautious, I decided to start with a 30-second stint and then wait a couple of days to see what happened. My knee pain didn't get worse. So I had another go in the garage, this time riding for 60 seconds. And waited another two days. Still, the pain held off. So I gradually upped the minutes until finally I was doing half an hour without any pain. I'd now experienced one of the many things Jane Trinca talked about: I had 'paced up' my activity with a graded exposure to exercise, and it worked. I was back on my bike. I had my freedom again, although my racing days, brief as they'd been, were over.

The experience would later ground my approach to the torn cartilage in my right knee. Teppo Järvinen's study, of course, featured prominently in the literature Fried sent my way after I stopped in to see him on that hot November day in 2019. After reading it, as you can imagine, I became quite sceptical about the benefits of a partial meniscectomy for my degenerative tear. So I started strengthening my quads, stretching my leg out with a homemade TheraBand substitute — an exercise not dissimilar to what Lawrie had done in the GLA:D program. And then I began a graded-exercise program, to see if I could get running again. On day one, I ran ten metres to the studio at the back of the garden. Day two, I rested. Day three, I ran to the studio and back again. Next day, rested. You get the picture. It wasn't long before I was running a couple of kilometres a week. What did I end up doing about the knee surgery? I cancelled it. I texted my surgeon that I would try to see it through without going under the knife. It's been three years now, and I never did get the op.

I want to make it clear, I'm no pain expert. Yes, I was a specialist medico for years, but, like many doctors, my understanding of the science of persistent pain was rudimentary at best. This book is a work of journalism, so I hope to have channelled the deep expertise of the researchers and of the people who have lived, and suffered, so long with pain. Yet, even though we've covered a lot of ground, the mountains of pain science range high and long, and, in truth, we've only just scratched the surface. Four decades on, Clifford Woolf is still a world-leading pain researcher. I read a recent review he co-authored of the challenges to developing effective analgesics, without crippling side effects, for acute and chronic pain. Even for the scientifically literate, it is almost unfathomably complex. It begins, however, with a clear description of the different pain types. There is acute pain from activation of nociceptors by intense heat or cold, mechanical force or chemical irritants, which is highly adaptive because it teaches us to avoid injury or to rest the injured part. There is pain from the inflammation that happens with infection, also adaptive because it gets you to rest the part, promoting healing. Then there are the maladaptive pains that, unfortunately, are simply not to our advantage. There is neuropathic pain, which, like Steve Olson's, comes from damaged nerves. And there is what has recently been coined 'nociplastic pain' — persistent pain in the absence of ongoing tissue damage, which is the subject of much of this book.

A book can't capture it all, but my own experience is a tangible example of how the research we have covered can help you get on top of pain, with minimal assistance from drugs or surgery. One important lesson takes me back, again, to Steve Olson. During hypnosis, he could turn his pain up and down with an imaginary dial. But, setting that example aside, the body is simply not a machine. For most of us, most of the time, persistent pain isn't something we can turn on — or, more importantly, off — much as we might want to. And there is at least one important reason why.

Not long after chatting with Olson, I began preparations to write the last chapter of this book. There was a sense of relief, but also a slight dread, because I knew I'd have to come up with a story, something meaningful that people would not only want to read, but could benefit from. That is a challenge. It's stressful. While I was in chapter-design mode, I was also dealing with yet another pain: in my left knee — again. But this time, it was exactly the same as what I'd experienced in my right knee, so I was confident that it was also a torn meniscus, and that it, too, would fade with time. Nevertheless, I thought I'd help it along with some exercises, so I drafted my faux TheraBand into service once more. Yet this brought on something unwelcome. My kneecap discomfort, a cardinal sign of patellofemoral-pain syndrome, reappeared. I hadn't felt it for years, and I started to worry. 'Is it back? Why did I do those exercises? Now I've brought the pain back.'

At the time, I was researching Vania Apkarian's work, and came across his passages about pain being a lifelong memory, and how the threshold for pain could be lowered by hypervigilance, where you're constantly on the lookout for it coming back. My anxiety increased. I was reminded of Lauren's incessant hunting for symptoms, and of Daniel Harvie telling me about the immense power of scary words, which our brain files away under the heading 'Danger'. I also thought, a little ruefully, about Henrik Vægter and his advice to find an exercise that didn't flare the injured part.

And then, amid all that stewing, came a funeral for a dear friend, held online because of the pandemic. Catherine died too young, from liver disease brought on, at least in part, by alcohol. I wanted to farewell her, but I had work to do, a schedule to keep. The funeral started and I was alone in my office, on my computer. I toyed with the idea of having a work window open on my browser, to chip away at the book prep while the funeral wound on. Catherine's partner opened with a long speech. Then a friend recited a poem called 'Over the Range' by Banjo Paterson. 'When we come to the final change, we shall meet with our loved ones gone before, to the beautiful country over the range,' she read. As I listened, I started paying more attention to the funeral service. Catherine's partner put up a slide show. There were pictures of her as a toddler, and when I first met her at a party in Melbourne in 1982, of her playing music in bands, and working at various jobs, sometimes wearing outfits I remembered, especially a dress with blue-and-black stripes. Pretty soon, I was sobbing away. But work still nipped at my heels and I got on with it again in the afternoon.

I started worrying more, troubled by the thought that I'd be trapped in a loop of watching for pain, watching for anxiety — and all of that making everything worse. I woke up that night in a sweat. The next day, I could feel the cold fingers of a low mood taking a grip. I felt awful. I had to lie down a lot. Small things set me off. My young boy's football had burst, but, even though it was deflated and broken, I couldn't put it in the bin, with all the memories of us learning to play together. I remembered my late friend Ricardo telling me once that we need to cry more. I cried more. I told myself I didn't need to be perfect, I could fail, be flawed, and it would be okay.

And amid all of it, my pain went away. It just disappeared, not through conscious effort but, perhaps like Lauren's experience in the car outside her doctor's surgery, from the emotional release.

It all reinforced how much pain is bound up in where our life is at. Stresses that trigger feeling low. And of course when you are stressed or depressed the body is just primed to sense threat. As so many people in this book have attested, pain is really the phenomenology, the feeling of your mind sensing danger. Ultimately, as Katja Wiech and Daniel Harvie made so clear, the notion that 'my body is in danger' is an inference, entangled with all the things that make threats, valid or not, seem real. From a scary X-ray report to a misshapen, broken body image that never got updated as healing progressed. So many of the messages I've heard from the people in this book converge on the idea that pain can be helped by changing what it means. By shifting pain away from being an overwhelming threat, indicative of catastrophic and irremediable damage to the body. As Lorimer Moseley put it, if we can just convince the brain that we don't need protection, there's a good chance it won't make pain.

I'm reminded of the medieval philosopher Peter Abelard, who anticipated the sentiments embraced a few centuries later in the Enlightenment. In Abelard's time, a prevailing dictum, attributed to Saint Anselm of Canterbury, could be summed up in the Latin *credo ut intelligam* — roughly, 'I must believe in order that I may understand.' With enough faith, the maxim tells us, the truth can be yours. Abelard begged to differ. His thinking was firmly rooted in the teachings of Aristotle, father of the syllogism and the use of logic in argument. So Abelard tipped the epigram on its head. 'I must understand in order that I may believe,' he said. 'By doubting we come to questioning, and by questioning we perceive the truth.'

Some people will change their beliefs through faith. We saw that in Katja Wiech's study that used images of the Virgin Mary. But many people need reason to believe, to base any shifted world view on propositions with good grounding in evidence. Pain science — whose modern era plausibly began with Clifford

Woolf's staggering discovery in London in 1983, but whose roots reach back through Wall and Melzack, to Beecher at the Anzio beachhead — offers a mountain of solid reasons to see many cases of persistent pain not as a sign of damage to the body, but as evidence of changes in the spinal cord and brain that conspire to keep us hurting long after the body has healed. The work of Tim Salomons shows how that information might ground a change of belief, through therapy, so powerful that it rewires the nervous system. And Tor Wager's work on pain reprocessing is strong evidence that belief change can, in some cases, relieve pain in its entirety. That could all be expedited with hypnosis, which, Mark Jensen's studies suggest, enhance the alpha and theta brainwaves that facilitate neuroplasticity and new learning. And then there is exercise. Kathleen Sluka has shown that exercise can reduce pain by an anti-inflammatory effect, but, as Christian Barton argued so convincingly, it can also act as evidence, presented in the silent courtroom of our daily experience, that the body is now safe, that it no longer needs protection.

When my knee pain came back, my friend died, and the challenge of that final book chapter loomed, I remembered a psychologist who had helped me nearly two decades earlier. I sent him an email. Given the demand on mental health professionals during the pandemic, I was surprised when he got back to me within hours. We did a Zoom meet, and he listened patiently, and then he offered some thoughts. Don't try to get rid of the anxiety, he said. You can tolerate it. If it comes, watch it like a train coming into the station. You're on the platform, but you don't need to get on the train, just watch it leave. It is the brain sending the signals that it is primed to when it thinks there's a threat. The signals will come and go, but they are not dangerous. Your aim is not to stop them, but a by-product of coming to accept them is that they will, gradually, diminish.

My anxiety did diminish, but not by any proclamation to banish it. Instead, it lessened through understanding, acceptance, and, using mindfulness, gently bringing my attention back to whatever I happened to be doing. It struck me that pain is similar in so many ways. The work of Apkarian and Wager, among others, shows how strongly pain is bound to the emotions, and so it makes sense that therapies that use mindfulness, or that aim at 'affective reappraisal', such as CBT and pain reprocessing, will help. Nociplastic pain, like the emotions, won't disappear with an edict from on high. It needs more work than that.

Naturally, we all have different ways of doing things. As a former tutor to medical students and doctors training to specialise in emergency medicine, I remember something I used to say. 'Do you know about the bowerbird?' I would ask them. Most didn't. 'The bowerbird collects blue things, and feathers its nest with them,' I told them, referencing a unique mating display of the bird, which is native to Australia and New Guinea. 'You need to collect medical things, and feather your nest with them,' I said, meaning they should hunt for, then collect, all the useful snippets of information they'd hear from their medical mentors.

It is my hope that you will become a bowerbird of pain things, collecting snippets here and there from experts and people who've been through it, and feather your nest with them, to become comfortable again. People with pain just want it to stop. Yet the journey is as important as the destination. The ride is challenging, scary, and feels too long, but so much of the experience is bound up in it. The very fact that you've taken the time to read this book, to collect your own snippets, is something I applaud you for, because listening to these experts and pain sufferers is at times not easy. It is, however, a sign you're on the search that is necessary to recover. Your journey will surely be unique and, when it's done, or nearly done, or you know that you'll be riding the tramcar on and off for a bit, I hope you'll feel up to sharing the wisdom you have gained, which will be like no one else's, with someone who is just beginning theirs.

Acknowledgements

I thank the people in this book who have selflessly made some of the most intimate, painful, and vulnerable moments of their lives open and transparent for the benefit of the rest of us. I thank the researchers, some of whom have been on their own pain journeys, for sharing their stories, and for their dedication, often lifelong, to the cause of ending chronic pain. If, in your darker moments, you despair at the way our world is heading, dwell on these people a while. Kal Fried calls for special mention. He's a warrior for the cause, who's helped me both personally as my physician and pain educator, and professionally as a guide to where the true source of pain often lies.

Many others helped along the way. Thank you, Lisa Hardwick, for collating data on the START program. Kenneth Goldberg, president of the Cleveland Heights Historical Society, along with Eric Silverman and Marian Morton, shared their deep local knowledge and stunning historical photos to nail the precise location of the Harold T. Clark Tennis Courts. Robert C. Nirschl, son of Dr Robert P. Nirschl, assisted with communication and provided a trove of helpful materials.

Thanks to Professor Dale Purves for permission to use his image of the Cornsweet illusion. Gratitude is owed also to Paul Visentini, Milly Bell, Professor James McAuley, Dr Martin Kroslak, Professor Bernard Morrey, Professor Robert Johnson, Professor Pierre Rainville, and Dr Yoni Ashar. I reserve special thanks for my editor at Scribe Publications, David Golding, whose superb corrections and suggestions taught me about the judicious use of questions, the proper limits of explanation, the value of brevity, and much more. Of course, deepest thanks, and love, to Luci Renault, and our children, Lydia, Adele, and Dylan, for keeping it all on track during challenging times.

Notes

Introduction. The Card Trick

The forces that pass through the knee would test Atlas: Harvard Medical School. 'Why Weight Matters When It Comes to Joint Pain'. 2019. Available: https://www.health.harvard.edu/pain/why-weight-matters-when-it-comes-to-joint-pain [Accessed 16 February 2022].

A good chunk of that compressive force is channelled through the meniscus: Fox AJS, Bedi A, Rodeo SA. 'The Basic Science of Human Knee Menisci: structure, composition, and function'. *Sports Health* 2012;4(4):340–51.

when you're running, the back, or 'posterior', of the medial meniscus: Luks HJ. 'Can I Make a Meniscus Tear Worse If I Run on It?' 2021. Available: https://www.howardluksmd.com/running-meniscus-tear/ [Accessed 16 February 2022].

it has been thoroughly mapped by MRI scans for over three decades: Reicher MA, et al. 'Meniscal Injuries: detection using MR imaging'. *Radiology* 1986;159(3).

fixing it with a procedure called a partial meniscectomy for even longer: McGinity JB, Geuss LF, Marvin RA. 'Partial or Total Meniscectomy: a comparative analysis'. *J Bone Joint Surg Am* 1977;59(6):763–6.

the torn flap of cartilage rubs against, and irritates, the glossy, translucent tissue that lines the knee joint, called the synovium: Roemer FW, et al. 'The Association Between Meniscal Damage of the Posterior Horns and Localized Posterior Synovitis Detected on T1-Weighted Contrast-Enhanced MRI': the MOST study'. *Semin Arthritis Rheum* 2013;42(6):573–81.

random bunch of people off the street in my age group — over 50 — one-third of them will have a torn meniscus: 60 per cent of those people with a torn meniscus, a clear majority, have no knee pain at all. Englund M, et al. 'Incidental Meniscal Findings on Knee MRI in Middle-Aged and Elderly Persons'. *N Engl J Med* 2008;359(11):1,108–15.

His website is peppered with blog posts: http://www.kalfried.com.au

It is called maladaptive neuroplasticity: Kuner R, Flor H. 'Structural Plasticity and Reorganisation in Chronic Pain'. *Nat Rev Neurosci* 2017;18(2):113.

In the US health system in 2016 alone, an eye-watering $380 billion was spent treating musculoskeletal disorders: Dieleman JL, et al. 'US Health Care Spending by Payer and Health Condition, 1996–2016'. *JAMA* 2020;323(9):863–84.

No fewer than one in five people live with chronic pain: Searing L. 'The Big Number: 50 million adults experience chronic pain'. *Washington Post* 21 October 2018.

overdose and suicide linked to opioid painkillers have been blamed for the first drop in US life expectancy in a century: DeWeerdt S. 'Tracing the US Opioid Crisis to its Roots'. *Nature* 2019;573(7,773):S10–S12.

Take spinal fusion for back pain. Rates doubled in the US in the decade to 2009 and the procedure generated costs exceeding $10 billion in 2015: Machado G, Lin C, Harris IA. 'Needless Treatments: spinal fusion surgery for lower back pain is costly and there's little evidence it'll work'. *The Conversation* 2018.

Yet one in six operations leads to complications, including nerve damage: Harris IA, et al. 'Lumbar Spine Fusion: what is the evidence?' *Intern Med J* 2018;48(12):1,430–4.

experts are lining up to say that, as a treatment for back pain, spinal fusion is as good as useless: ibid.

a learning module for the Australasian College of Sport and Exercise Physicians: Fried K. '"Protection" Neurobiology — a Key Piece of the Pain Puzzle'. SEM Academy. 2020. Available: https://semacademy.org/collections/featured-modules/products/pain-management-1 [Accessed 16 February 2022].

1. Who Got the Dux?

meralgia paraesthetica. An entrapped nerve produces a palette of pain: Cheatham SW, Kolber MJ, Salamh PA. 'Meralgia Paresthetica: a review of the literature'. *Int J Sports Phys Ther* 2013;8(6):883–93.

a small percentage of sufferers get it on both sides: Ivins GK. 'Meralgia Paresthetica, the Elusive Diagnosis: clinical experience with 14 adult patients'. *Ann Surg* 2000;232(2):281–6.

Woolf quizzed him. 'Why are you doing that? How does it work?': Zapol WM. 'Clifford J. Woolf, MB, BCh, PhD: recipient of the 2004 Excellence in Research Award'. *Anesthesiology* 2004;101(4):820–3.

As a medical student at Oxford in the 1940s, he chaired the socialist club: 'Patrick D. Wall.' In: Squire LR, editor. *The History of Neuroscience in Autobiography*. San Diego, California: Academic Press, 2001.

TRIO: The Revolting Intellectuals Organization: Wall PD. *TRIO: The Revolting Intellectuals Organization*. CN Potter, 1965.

Wall, along with Canadian psychologist Ronald Melzack, had in 1965 published a radical paper: Melzack R, Wall PD. 'Pain Mechanisms: a new theory'. *Science* 1965;150(3,699):971–9.

In his book Treatise on Man, *published in 1633, Descartes includes the bizarre image of a person*: Descartes R. *The World and Other Writings*. Cambridge, New York: Cambridge University Press, 1998.

just as when you pull on one end of a cord you cause a bell hanging at the other end to ring at the same time: ibid.: p. 117.

'that was all there was to it,' said Wall in a 1999 interview: Mitchell N. 'Pat Wall: pain man of the century'. ABC Radio, 2003.

One 'husky 19-year-old soldier', Beecher recalled later: Beecher HK. 'Pain in Men Wounded in Battle'. *Ann Surg* 1946;123(1):96–105.

Three-quarters of badly wounded men, although they have received no morphine: ibid.

Beecher selected 150 civilians having these and similar types of surgery: Beecher HK. 'Relationship of Significance of Wound to Pain Experienced'. *JAMA* 1956;161(17):1,609–13.

In humans, there are thousands of these motor neurons exiting the spinal cord: Tomlinson BE, Irving D. 'The Numbers of Limb Motor Neurons in the Human Lumbosacral Cord Throughout Life'. *J Neurol Sci* 1977;34(2):213–19.

enter the spinal cord and block the flexion withdrawal reflex: Baars JH, et al. 'Effects of Sevoflurane and Propofol on the Nociceptive Withdrawal Reflex and on the H Reflex'. *Anesthesiology* 2009;111(1):72–81.

In the end, Nature *published the paper*: Woolf CJ. 'Evidence for a Central Component of Post-Injury Pain Hypersensitivity'. *Nature* 1983;306(5,944):686–8.

Pain Revolution: http://www.painrevolution.org.

dubbed 'one of the most creative pain researchers alive': Doidge N. *The Brain's Way of Healing: remarkable discoveries and recoveries from the frontiers of neuroplasticity*. Brunswick, Victoria: Scribe Publications, 2015: p. 23.

'belly breathing', also called diaphragmatic breathing, which can ease pain: White J. 'The Effect of Diaphragmatic Breathing on Pain Pressure Threshold in Patients with Central Sensitization'. ProQuest Dissertations Publishing, 2019.

2. The Rabbit Hole of Who You Are

The chances were just 5 per cent, but that's exactly what had happened: Kao F-C, et al. 'Short-Term and Long-Term Revision Rates After Lumbar Spine Discectomy Versus Laminectomy: a population-based cohort study'. *BMJ Open* 2018;8(7):e021028.

injected a glucose solution into the space around the spinal cord: Maniquis-Smigel L, et al. 'Short-Term Analgesic Effects of 5% Dextrose Epidural Injections for Chronic Low Back Pain: a randomized controlled trial'. *Anesth Pain Med* 2017;7(1):e42550.

radiofrequency ablation, which uses electrical current to heat up and knock out nerves: De Vivo AE, et al. 'Intra-Osseous Basivertebral Nerve Radiofrequency Ablation (BVA) for the Treatment of Vertebrogenic Chronic Low Back Pain'. *Neuroradiology* 2020;63(5):809–15.

stem cells taken from the marrow of his hip bone: Urits I, et al. 'Stem Cell Therapies for Treatment of Discogenic Low Back Pain: a comprehensive review'. *Curr Pain Headache Rep* 2019;23(9):1–12.

their flagship program, called START: START stands for 'Selected Targets of Activity ReTraining' (Jane Trinca, personal communication).

The big one is something called opioid-induced hyperalgesia: Roeckel L-A, et al. 'Opioid-Induced Hyperalgesia: cellular and molecular mechanisms'. *Neuroscience* 2016;338:160–82.

One review found a fifth of fusions needed reoperation within five years: Tobert DG, et al. 'Adjacent Segment Disease in the Cervical and Lumbar Spine'. *Clin Spine Surg* 2017;30(3):94–101.

spinal fusion should not be done in people without signs of nerve compression: ANZCA. Do Not Refer Axial Lower Lumbar Back Pain for Spinal Fusion Surgery'. Choosing Wisely Australia, 2018.

Here is how Moseley describes it in his book Painful Yarns: Moseley GL. *Painful Yarns: metaphors & stories to help understand the biology of pain*. Canberra, Australia: Dancing Giraffe Press, 2007: p. 63.

So she set about surveying a bunch of pain professionals: Madden VJ, Moseley GL. 'Do Clinicians Think that Pain Can Be a Classically Conditioned Response to a Non-Noxious Stimulus?' *Man Ther* 2015;22:165–73.

researchers spread the word they were on the hunt for study participants: Madden VJ, et al. 'Pain by Association?: experimental modulation of human pain thresholds using classical conditioning'. *J Pain* 2016;17(10):1,105–15.

Pavlov used a metronome, whose regular ticking worked a treat: Black SL. 'Pavlov's Dogs: for whom the bell rarely tolled'. *Curr Biol* 2003;13(11):R426.

Bells and whistles, for example, got the canine juices flowing, too: Razran G. 'Stimulus Generalization of Conditioned Responses'. *Psychol Bull* 1949;46(5):337–65.

The phenomenon is called stimulus generalisation: Hughes JW. 'Stimulus Generalization'. *Encyclopedia of Clinical Neuropsychology*. New York: Springer, 2011.

groundbreaking paper written with Dutch researcher Johan Vlaeyen in 2015: Moseley GL, Vlaeyen JWS. 'Beyond Nociception: the imprecision hypothesis of chronic pain'. *Pain* 2015;156(1):35–8.

Most people will feel two points. But the palm of your hand is less sensitive: Mancini F, et al. 'Whole-body Mapping of Spatial Acuity for Pain and Touch'. *Ann Neurol* 2014;75(6):917–24.

people with chronic lower-back pain needed the pins to be a full 12 millimetres wider: Adamczyk W, Luedtke K, Saulicz E. 'Lumbar Tactile Acuity in Patients with Low Back Pain and Healthy Controls: systematic review and meta-analysis'. *Clin J Pain* 2018;34(1):82–94.

trigger an explosive release of endorphins, the body's inner painkillers: Peciña M, Zubieta JK. 'Molecular Mechanisms of Placebo Responses in Humans'. *Mol Psychiatry* 2015;20(4):416–23.

they give up trying to emulate the real treatment and go for something completely different: Machado LAC, et al. 'Imperfect Placebos Are Common in Low Back Pain Trials: a systematic review of the literature'. *Eur Spine J* 2008;17(7):889–904.

enlisting 276 adults who'd put up with lower-back pain for at least 12 weeks: Bagg MK, et al. 'The RESOLVE Trial for People with Chronic Low Back Pain: protocol for a randomised clinical trial'. *J Physiother* 2016;63(1):47–8.

a study that got people with back pain to look at a series of fairly repetitive photos: Bray H, Moseley GL. 'Disrupted Working Body Schema of the Trunk in People with Back Pain'. *Br J Sports Med* 2011;45(3):168–73.

used magnets to stimulate the brain's motor cortex and measured the response: Schabrun SM, Elgueta-Cancino EL, Hodges PW. 'Smudging of the Motor Cortex Is Related to the Severity of Low Back Pain'. *Spine* 2017;42(15):1,172–8.

muscles stabilise the spine and make it harder to use: Hodges PW, et al. 'Diverse Role of Biological Plasticity in Low Back Pain and Its Impact on Sensorimotor Control of the Spine'. *J Orthop Sports Phys Ther* 2019;49(6):389–401.

you need a flexible spine to share load evenly across the discs: Brumagne S, et al. 'Neuroplasticity of Sensorimotor Control in Low Back Pain'. *J Orthop Sports Phys Ther* 2019;49(6):402–14. Hodges PW, et al. 'New Insight into Motor Adaptation to Pain Revealed by a Combination of Modelling and Empirical Approaches'. *Eur J Pain* 2013;17(8):1,138–46.

They were given feedback to ensure their back movements became fluid: Bagg MK, et al. 'The RESOLVE Trial for People with Chronic Low Back Pain: protocol for a randomised clinical trial'. *J Physiother* 2016;63(1):47–8.

the 'real' treatment group had a full point reduction: Bagg MK, et al. 'Effect of Graded Sensorimotor Retraining on Pain Intensity in Patients with Chronic Low Back Pain'. *JAMA* 2022;328(5):430–9.

an emerging concept called bioplasticity: Petrova S. 'Trust Me I'm An Expert: the science of pain'. *The Conversation*, 2018.

3. An Old, Rusty Robot

they could turn their heads seven degrees further before getting any pain: Harvie DS, et al. 'Bogus Visual Feedback Alters Onset of Movement-Evoked Pain in People with Neck Pain'. *Psychol Sci* 2015;26(4):385–92.

It's as if the skin of my arm has been ripped off: Nortvedt F, Engelsrud G. '"Imprisoned" in Pain: analyzing personal experiences of phantom pain'. *Med Health Care Philos* 2014;17(4):599–608.

Numerous studies suggest this can help phantom-limb pain: Finn SB, et al. 'A Randomized, Controlled Trial of Mirror Therapy for Upper Extremity Phantom Limb Pain in Male Amputees'. *Front Neurol* 2017;8:267.

One macabre study had participants in VR goggles rest their hand on a desktop: González-Franco M, et al. 'A Threat to a Virtual Hand Elicits Motor Cortex Activation'. *Exp Brain Res* 2013;232(3):875–87.

a pain-management course at Gold Coast University Hospital: Thomson H, et al. 'Identifying Psychosocial Characteristics That Predict Outcome to the UPLIFT Programme for People with Persistent Back Pain: protocol for a prospective cohort study'. *BMJ Open* 2019;9(8):e028747.

documented Cole's amazing results in a paper published in 2020: Harvie DS, et al. 'Virtual Reality Body Image Training for Chronic Low Back Pain: a single case report'. *Front Virtual Real* 2020;1(13).

a stunning portrait of a young woman, eyebrows raised haughtily: This is *Portrait of a Young Girl* by the Dutch artist Petrus Christus.

The pair enlisted the help of 12 young Catholics from Oxford: Wiech K, et al. 'An fMRI Study Measuring Analgesia Enhanced by Religion as a Belief System'. *Pain* 2008;139(2):467–76.

it's known to be active when we rejig our understanding of things in order to feel better: Marques LM, Morello LYN, Boggio PS. 'Ventrolateral but Not Dorsolateral Prefrontal Cortex tDCS Effectively Impact Emotion Reappraisal: effects on emotional experience and interbeat interval'. *Sci Rep* 2018;8(1):15,295.

can modify nerve activity in the spinal cord to tamp down pain signals: Wager TD, Atlas LY. 'The Neuroscience of Placebo Effects: connecting context, learning and health'. *Nat Rev Neurosci* 2015;16(7):403–18.

Did you read it as 'if I' instead of 'I if'? If you did, don't worry — you're not alone: Howlett C. 'Predictive Processing: a potential theory for persistent pain, and the power of discrepancy in facilitating change'. 2019. Available: https://www.noigroup.com/noijam/predictive-processing-a-potential-theory-for-persistent-pain-and-the-power-of-discrepancy-in-facilitating-change/ [Accessed 17 February 2022].

we don't waste precious energy deciphering outliers such as 'I if': Jepma M, et al. 'Behavioural and Neural Evidence for Self-Reinforcing Expectancy Effects on Pain'. *Nat Hum Behav* 2018;2(11):838–55.

This is called predictive processing: Wiech K. 'Biased Perception and Learning in Pain'. *Nat Hum Behav* 2018;2(11):804–5.

They enlisted the help of 22 healthy people: Bingel U, et al. 'The Effect of Treatment Expectation on Drug Efficacy: imaging the analgesic benefit of the opioid remifentanil'. *Sci Transl Med* 2011;3(70):70ra14.

working with a crack team of scientists at the University of Toronto: Salomons TV, et al. 'A Brief Cognitive-Behavioural Intervention for Pain Reduces Secondary Hyperalgesia'. *Pain* 2014;155(8):1,446–52.

in the midst of massively upscaling this smaller study: Salomons TV. 'Modulatory Capacity: biomarkers for pain sensitization and response to psychological treatment for pain'. 2022. Available: https://gtr.ukri.org/projects?ref=MR%2FR005656%2F1 [Accessed 17 February 2022].

4. Adherence Is Critical

The resulting picture is oddly intoxicating: https://www.kathleenslukaart.com/um5twninc0y6awe1vg88h7uz3cv1g6

You can see Sluka's prolific and stunning array of art on her website: https://www.kathleenslukaart.com

The research showed that you had to exercise hard before endorphins got pumping: Goldfarb AH, Jamurtas AZ. 'β-Endorphin Response to Exercise: an update'. *Sports Med* 1997;24(1):8–16.

the rats' pain thresholds had returned to normal: Bement MKH, Sluka KA. 'Low-Intensity Exercise Reverses Chronic Muscle Pain in the Rat in a Naloxone-Dependent Manner'. *Arch Phys Med Rehabil* 2005;86(9):1,736–40.

One of Sluka's paintings features a bunch of pompoms: https://www.kathleenslukaart.com/shadow-self-portrait

in a case of unwitting suicide, the macrophages promptly die: Van Rooijen N. 'The Liposome-Mediated Macrophage "suicide" Technique'. *J Immunol Methods* 1989;124(1):1–6.

Sluka injected liposomes into the rats' calves to lay waste to all the macrophages: Gong W-Y, et al. 'Resident Macrophages in Muscle Contribute to Development of Hyperalgesia in a Mouse Model of Non-Inflammatory Muscle Pain'. *J Pain* 2016;17(10):1,081–94.

tweaked the injection, making it slightly less acidic: Gregory NS, et al. 'An Overview of Animal Models of Pain: disease models and outcome measures'. *J Pain* 2013;14(11):1255–69.

Fatiguing exercise was prompting macrophages to spit out pain-pinging proteins: Gong W-Y, et al. 'Resident Macrophages in Muscle Contribute to Development of Hyperalgesia in a Mouse Model of Non-Inflammatory Muscle Pain'. *J Pain* 2016;17(10):1,081–94.

So Sluka put running wheels in the animals' cages: Sluka KA, et al. 'Regular Physical Activity Prevents Development of Chronic Pain and Activation of Central Neurons'. *J Appl Physiol* 2013;114(6):725–33.

their muscles were now packed with M2 macrophages: Leung A, et al. 'Regular Physical Activity Prevents Chronic Pain by Altering Resident Muscle Macrophage Phenotype and Increasing IL-10 in Mice'. *Pain* 2016;157(1):70–9.

Work absenteeism dropped 34 per cent: Auken G. 'There Is Labor in the Long-Term Sick'. 2007. Available: https://www-information-dk.translate.goog/2007/07/arbejdskraft-langtidssyge?_x_tr_sl=da&_x_tr_tl=en&_x_tr_hl=en&_x_tr_pto=sc [Accessed 17 February 2022].

questioned more than 46,000 Norwegians and found that 29 per cent had chronic pain: 'Associations Between Recreational Exercise and

Chronic Pain in the General Population: evidence from the HUNT 3 study'. *Nat Rev Neurol* 2011;7(7):357.

recruited 96 adults with lower-back pain: Vægter HB, et al. 'Impaired Exercise-Induced Hypoalgesia in Individuals Reporting an Increase in Low Back Pain During Acute Exercise'. *Eur J Pain* 2021;25(5):1,053–63.

The study, published in 2020, was called 'Power of Words': Vægter HB, et al. 'Power of Words: influence of preexercise information on hypoalgesia after exercise — randomized controlled trial'. *Med Sci Sports Exerc* 2020;52(11):2,373–9.

Barton has recently put the finishing touches on the 2020 annual report: GLA:D Australia. *Annual Report 2020*. Available: https://gladaustralia.com.au/annual-reports/.

asked his then colleague Geoffrey Vaupel to do an arthroscopy on his knee: Dye SF, Vaupel GL, Dye CC. 'Conscious Neurosensory Mapping of the Internal Structures of the Human Knee Without Intraarticular Anesthesia'. *Am J Sports Med* 1998;26(6):773–7.

Bruising of the underlying bone, which can show up on an MRI, is linked to pain: Singh VV, et al. 'Clinical and Pathophysiologic Significance of MRI Identified Bone Marrow Lesions Associated with Knee Osteoarthritis'. *Arch Bone Jt Surg* 2019;7(3):211–19.

it wasn't by reducing force through the knee or altering biomechanics: DeVita P, et al. 'Quadriceps-Strengthening Exercise and Quadriceps and Knee Biomechanics During Walking in Knee Osteoarthritis: a two-centre randomized controlled trial'. *Clin Biomech* 2018;59:199–206.

Eduard Alentorn-Geli led a team that scoured studies about osteoarthritis in runners: Alentorn-Geli E, et al. 'The Association of Recreational and Competitive Running with Hip and Knee Osteoarthritis: a systematic review and meta-analysis'. *J Orthop Sports Phys Ther* 2017;47(6):373–90.

it doesn't work for roughly one in five people: Price AJ, et al. 'Knee Replacement'. *Lancet* 2018;392(10,158):1,672–82.

The main reason to get a knee replacement is to relieve pain: Medical Advisory Secretariat. 'Total Knee Replacement: an evidence-based analysis'. *Ont Health Technol Assess Ser* 2005;5(9):1–51.

5. Passengers on the Bus

the backhand delivers a force of around 70 newtons: Riek S, Chapman AE, Milner T. 'A Simulation of Muscle Force and Internal Kinematics of Extensor Carpi Radialis Brevis During Backhand Tennis Stroke: implications for injury'. *Clin Biomech* 1999;14(7):477–83.

if you scan the ranks at the top levels of the game, rates of tennis elbow don't go up: De Smedt T, et al. 'Lateral Epicondylitis in Tennis: update on aetiology, biomechanics and treatment'. *Br J Sports Med* 2007;41(11):816–19.

Around four out of five people with tennis elbow got better within a year: American Academy of Orthopaedic Surgeons. 'Tennis Elbow (Lateral Epicondylitis)'. 2020. Available: https://orthoinfo.aaos.org/en/diseases — conditions/tennis-elbow-lateral-epicondylitis/.

Jim Kaplan came and interviewed him for a profile in Sports Illustrated: Kaplan J. 'The Doc Who Tells You What's Up'. *Sports Illustrated* 1976.

worked about 50 per cent of the time, as often as a favourable coin toss: Robert Nirschl, personal communication.

When he makes that first incision through the skin next to the elbow: https://www.youtube.com/watch?v=bTBdaUGDFzE

tendons that looked like fibreboard, a bunch of cells sprayed together and loosely compacted: Kraushaar BS, Nirschl RP. 'Tendinosis of the Elbow (Tennis Elbow): clinical features and findings of histological, immunohistochemical, and electron microscopy studies'. *J Bone Joint Surg Am* 1999;81(2):259–78.

publish his results in a series of 88 surgeries: Nirschl RP, Pettrone FA. 'Tennis Elbow: the surgical treatment of lateral epicondylitis'. *J Bone Joint Surg Am* 1979;61(6):832–9.

overturned long-held theories about food poisoning: Murrell TG, Roth L. 'Necrotizing Jejunitis: a newly discovered disease in the highlands of New Guinea'. *The Medical journal of Australia* 1963;50(1)(3):61–9.

led to some consternation among those still invested in the old theories: Walker PD. 'Obituary of Timothy Murrell'. *PNG Med J* 2003. 46(1–2):87–9.

He gave the impression that he didn't really think the surgery was that worthwhile: It's worth noting that Bernard Morrey is puzzled by that recollection. 'I'm surprised that I had given a lecture ... where anybody thought I was saying the Nirschl procedure doesn't work,' Morrey told me. 'I think that I expressed some contradiction in the literature in that it was unclear exactly what the role of intervention was, but it did appear to be beneficial in many people's hands.'

using a device Murrell helped design and build: Paoloni JA, Appleyard RC, Murrell GAC. 'The Orthopaedic Research Institute–Tennis Elbow Testing System: a modified chair pick-up test—interrater and intrarater reliability testing and validity for monitoring lateral epicondylosis'. *J Shoulder Elbow Surg* 2004;13(1):72–7.

there were, quite simply, no differences between the groups: Kroslak M, Murrell GAC. 'Surgical Treatment of Lateral Epicondylitis: a prospective, randomized, double-blinded, placebo-controlled clinical trial'. *Am J Sports Med* 2018;46(5):1,106–13.

Something called common factors: Wampold BE. 'How Important Are the Common Factors in Psychotherapy?: an update'. *World Psychiatry* 2015;14(3):270–7.

One really important common factor is allegiance: Hollon SD. 'Allegiance Effects in Treatment Research: a commentary'. *Clinical Psychol* 1999;6(1):107–12.

Chang's extraordinary study, published in Nature Human Behaviour: Chen P-HA, et al. 'Socially Transmitted Placebo Effects'. *Nat Hum Behav* 2019;3(12):1,295–305.

Pilots use 'we' and 'our' more often than their subordinate first officers: Kacewicz E, et al. 'Pronoun Use Reflects Standings in Social Hierarchies'. *J Lang Soc Psychol* 2014;33(2):125–43.

James W. Pennebaker, from the University of Texas at Austin, led a study: ibid.

the same traits are found to predict a placebo response: Kaptchuk TJ, et al. 'Components of Placebo Effect: randomised controlled trial in patients with irritable bowel syndrome'. *BMJ* 2008;336(7,651):999–1,003.

Each year, 650,000 of the procedures were done in the US alone: Moseley JB, et al. 'A Controlled Trial of Arthroscopic Surgery for Osteoarthritis of the Knee'. *N Engl J Med* 2002;347(2):81–8.

people who got the placebo operation did just as well: ibid.

Having a torn meniscus, Englund had shown, was totally compatible with having no pain: Englund M, et al. 'Incidental Meniscal Findings on

Knee MRI in Middle-Aged and Elderly Persons'. *N Engl J Med* 2008;359(11):1,108–15.

people who got the real surgery and the fake surgery did equally well: Sihvonen R, et al. 'Arthroscopic Partial Meniscectomy Versus Sham Surgery for a Degenerative Meniscal Tear'. *N Engl J Med* 2013;369(26):2,515–24.

A group of surgeons claimed The New England Journal of Medicine *was biased*: Rossi MJ, et al. 'Could the *New England Journal of Medicine* Be Biased Against Arthroscopic Knee Surgery?' *Arthroscopy* 2014;30(5):536–7.

individuals born and raised in Finland cannot be extrapolated to the rest of the world: Sochacki KR, et al. 'Sham Surgery Studies in Orthopaedic Surgery May Just Be a Sham: a systematic review of randomized placebo-controlled trials'. *Arthroscopy* 2020;36(10):2750–62.e2.

APM rates remain high in the US, but they are falling in Finland: Ardern CL, et al. 'When Taking a Step Back Is a Veritable Leap Forward. Reversing Decades of Arthroscopy for Managing Joint Pain: five reasons that could explain declining rates of common arthroscopic surgeries'. *Br J Sports Med* 2020;54(22):1,312–13.

Am I going to be better with the surgery compared to not having the surgery?: https://www.youtube.com/watch?v=jpqRngDEp_E

results of the five-year follow-up on FIDELITY *participants*: Sihvonen R, et al. 'Arthroscopic Partial Meniscectomy for a Degenerative Meniscus Tear: a 5 year follow-up of the placebo-surgery controlled FIDELITY (Finnish Degenerative Meniscus Lesion Study) trial'. *Br J Sports Med* 2020;54(22):1,332–9.

6. The Safety Matrix

using hypnosis for PICC *lines and a procedure called sclerosis*: 'Efficacy of Hypnoanalgesia by a Radiologist Technologist in Children with Cutaneous Angioma Treated with Sclerosis in Interventional Radiology'. https://ichgcp.net/clinical-trials-registry/NCT03674346

a procedure with local anaesthetic that, according to one recent study, requires sedation in nearly two-thirds of cases: Rainey SC, et al. 'Development of a Pediatric PICC Team Under an Existing Sedation Service: a 5-year experience'. *Clin Med Insights Pediatr* 2019;13:1179556519884040.

Try to pose for yourself this task: not to think of a polar bear: Dostoyevsky F. *Winter Notes on Summer Impressions*. Evanston, Illinois: Northwestern University Press, 1997: p. 49.

once introduced, ejecting images of a white bear from one's mind is well-nigh impossible: Wegner DM, et al. 'Paradoxical Effects of Thought Suppression'. *J Pers Soc Psychol* 1987;53(1):5–13.

Negating words … fail to mitigate the effects of negatively valued words: Cyna AM, Lang EV. 'How Words Hurt'. In: Cyna AM, et al., editors. *Handbook of Communication in Anaesthesia and Critical Care: a practical guide to exploring the art*. Oxford: Oxford University Press, 2010.

L'Écuyer's team had enough data to present the results of their pilot: Fortin M, Fortin V, L'Écuyer J. 'Hypnosis in Medical Imaging'. PowerPoint presentation, November 2019. Vicky Fortin, personal communication.

A recent Australian trial of hypnosis for people having burns dressed: Chester SJ, et al. 'Efficacy of Hypnosis on Pain, Wound-Healing, Anxiety, and Stress in Children with Acute Burn Injuries: a randomized controlled trial'. *Pain* 2018;159(9):1,790–801.

issued a 'strong recommendation' for 'the use of hypnosis for all needle procedures': Loeffen EAH, et al. 'Reducing Pain and Distress Related to Needle Procedures in Children with Cancer: a clinical practice guideline'. *Eur J Cancer* 2020;131:53–67.

Testosterone replacement sometimes helps with pain: Basaria S, et al. 'Effects of Testosterone Replacement in Men with Opioid-Induced Androgen Deficiency: a randomized controlled trial'. *Pain* 2015;156(2):280–88.

a full two-point reduction on a ten-point pain scale, now maintained for five years: McKittrick ML, et al. 'Hypnosis for Refractory Severe Neuropathic Pain: a case study'. *Am J Clin Hypn* 2020;63(1):28–35.

clinically meaningful reductions in pain: Thompson T, et al. 'The Effectiveness of Hypnosis for Pain Relief: a systematic review and meta-analysis of 85 controlled experimental trials'. *Neurosci Biobehav Rev* 2019;99:298–310.

Alphonse Piché, who is eulogised in one biographical entry: Hill RG. 'Piche, Alphonse'. *Biological Dictionary of Architects in Canada 1800–1950*. 2009. Available: http://dictionaryofarchitectsincanada.org/node/1393 [Accessed 18 February 2022].

Desmarteaux and the team have just finished their analysis of the data: Desmarteaux C, et al. 'Brain Responses to Hypnotic Verbal Suggestions Predict Pain Modulation'. *Front Pain Res* 2021;2:757384.

great neurologist Charcot was delivering one of his famous Tuesday Lessons: Gelfand T. 'Tuesdays at the Salpêtrière'. *Bull Hist Med* 1989;63(1):132–6.

you can reverse placebo analgesia with the drug naloxone: Benedetti F. 'The Opposite Effects of the Opiate Antagonist Naloxone and the Cholecystokinin Antagonist Proglumide on Placebo Analgesia'. *Pain* 1996;64(3):535–43.

But naloxone does not reverse the pain relief produced by hypnosis: Price DD, Rainville P. 'Hypnotic Analgesia'. In: Genhart GF, Schmidt RF, editors. *Encyclopedia of Pain*. Heidelberg, Germany: Springer, 2013: pp. 1,537–42.

reviewed the increasing number of studies on CBT: Williams ACdC, Eccleston C, Morley S. 'Psychological Therapies for the Management of Chronic Pain (Excluding Headache) in Adults'. *Cochrane Database Syst Rev* 2012;11(11):CD007407.

They had combined CBT with hypnosis to treat an illness: Kirsch I, Montgomery G, Sapirstein G. 'Hypnosis as an Adjunct to Cognitive-Behavioral Psychotherapy: a meta-analysis'. *J Consult Clin Psychol* 1995;63(2):214–20.

added a hypnotic induction to CBT for people in chronic pain: Edelson J, Fitzpatrick JL. 'A Comparison of Cognitive Behavioral and Hypnotic Treatments of Chronic Pain'. *J Clin Psychol* 1989;45(2):316–23.

education alone did very little to reduce the average pain intensity of those people with MS: Jensen MP, et al. 'Effects of Self-Hypnosis Training and Cognitive Restructuring on Daily Pain Intensity and Catastrophizing in Individuals with Multiple Sclerosis and Chronic Pain'. *Int J Clin Exp Hypn* 2011;59(1):45–63.

In 2020, the team published their findings: Jensen MP, et al. 'Effects of Hypnosis, Cognitive Therapy, Hypnotic Cognitive Therapy, and Pain Education in Adults with Chronic Pain: a randomized clinical trial'. *Pain* 2020;161(10):2,284–98.

slow-wave oscillations are incredibly important to new learning: Sigala R, et al. 'The Role of Alpha-Rhythm States in Perceptual Learning:

insights from experiments and computational models'. *Front Comput Neurosci* 2014;8:36.

Catastrophic thinking, they found, descreased in all four treatement groups: Jensen MP, et al. 'Pain-Related Beliefs, Cognitive Processes, and Electroencephalography Band Power as Predictors and Mediators of the Effects of Psychological Chronic Pain Interventions'. *Pain* 2021;162(7):2,036–50.

Hypnotic susceptibility has a genetic component: Rominger C, et al. 'Carriers of the COMT Met/Met Allele Have Higher Degrees of Hypnotizability, Provided That They Have Good Attentional Control: a case of gene-trait interaction'. *Int J Clin Exp Hypn* 2014;62(4):455–82.

what you expect to happen in hypnosis, while a factor: Kirsch I. 'Hypnosis and Placebos: response expectancy as a mediator of suggestion effects'. *An de Psicol* 1999;15(1).

bears less relation to whether it will work for you: Benham G, et al. 'Self-Fulfilling Prophecy and Hypnotic Response Are Not the Same Thing'. *J Pers Soc Psychol* 1998;75(6):1,604–13.

On one estimate, around 10–15 per cent of people are highly susceptible: Santarcangelo EL, Consoli S. 'Complex Role of Hypnotizability in the Cognitive Control of Pain'. *Front Psychol* 2018;9:2,272.

7. The Lens of Mindset

let the person report pain by simply moving those digits: Apkarian AV, et al. 'Imaging the Pain of Low Back Pain: functional magnetic resonance imaging in combination with monitoring subjective pain perception allows the study of clinical pain states'. *Neurosci Lett* 2001;299(1):57–60.

It had shifted to a completely different set of regions: Hashmi JA, et al. 'Shape Shifting Pain: chronification of back pain shifts brain representation from nociceptive to emotional circuits'. *Brain* 2013;136(Pt 9):2,751–68.

It was a series of 24 cylindrical boxes: Heron WT, Skinner BF. 'An Apparatus for the Study of Animal Behavior'. *Psychol Rec* 1939;3:165.

The amygdala is also key to the acquisition of fear memories: Barroso J, Branco P, Apkarian AV. 'Brain Mechanisms of Chronic Pain: critical role of translational approach'. *Transl Res* 2021;238:76–89.

in persistent pain, the hippocampus undergoes subtle changes in shape: Berger SE, et al. 'Hippocampal Morphology Mediates Biased Memories of Chronic Pain'. *NeuroImage* 2018;166:86–98.

the role of the hippocampus, like so much about pain, is complicated: Kim WB, Cho J-H. 'Encoding of Contextual Fear Memory in Hippocampal-Amygdala Circuit'. *Nat Commun* 2020;11(1):1,382–22.

the hippocampus stores fear memory for places: Vania Apkarian, personal communication.

chronic pain is not 'the temporal extension of acute pain': Barroso J, Branco P, Apkarian AV. 'Brain Mechanisms of Chronic Pain: critical role of translational approach'. *Transl Res* 2021;238:76–89.

avoiding *things that make pain worse becomes a reward in and of itself*: Budygin EA, et al. 'Aversive Stimulus Differentially Triggers Subsecond Dopamine Release in Reward Regions'. *Neuroscience* 2011;201:331–7.

the ones most likely to get persistent pain: Apkarian AV, Baliki MN, Farmer MA. 'Predicting Transition to Chronic Pain'. *Curr Opin Neurol* 2013;26(4):360–7.

the algorithm, in an extraordinary result, had delivered: Wager TD, et al. 'An fMRI-Based Neurologic Signature of Physical Pain'. *N Engl J Med* 2013;368(15):1,388–97.

a striking set of images of the brain, studded with blue dots: Wager TD, Atlas LY. 'The Neuroscience of Placebo Effects: connecting context, learning and health'. *Nat Rev Neurosci* 2015;16(7):403–18.

stimulating the vmPFC can shut down the gate on incoming pain signals: Dale J, et al. 'Scaling Up Cortical Control Inhibits Pain'. *Cell Rep* 2018;23(5):1,301–13. Lee M, et al. 'Activation of Corticostriatal Circuitry Relieves Chronic Neuropathic Pain. *J Neurosci* 2015;35(13):5,247–59.

to use their imagination to make the pain seem worse: Choong-Wan W, et al. 'Distinct Brain Systems Mediate the Effects of Nociceptive Input and Self-Regulation on Pain'. *PLoS biology* 2015;13(1):e1002036.

it's the ventromedial prefrontal circuit, not classic nociceptive circuits: Wager TD. 'Why Do Some of My Patients Have So Much More Pain Than I Think They Should?' Lecture to the 2018 ANZCA Annual Scientific Meeting. Available: https://vimeo.com/268725793 [Accessed 18 February 2022].

The default-mode network is critical when we discern value: Ashar YK, Chang LJ, Wager TD. 'Brain Mechanisms of the Placebo Effect: an affective appraisal account'. *Annu Rev Clin Psychol* 2017;13(1):73–98.

a number of studies show it can be as effective as a 'deceptive' placebo: Kaptchuk TJ. 'Open-Label Placebo: reflections on a research agenda'. *Perspect Biol Med* 2018;61(3):311–34.

pain reappraisals are also made real by emotions — positive ones: Ashar YK, Chang LJ, Wager TD. 'Brain Mechanisms of the Placebo Effect: an affective appraisal account'. *Annu Rev Clin Psychol* 2017;13(1):73–98.

a stance of detached, non-judgemental watching, which can de-escalate things: Garland EL, et al. 'Randomized Controlled Trial of Brief Mindfulness Training and Hypnotic Suggestion for Acute Pain Relief in the Hospital Setting'. *J Gen Intern Med* 2017;32(10):1,106–13.

The University of Colorado back-pain study, published in JAMA Psychiatry: Ashar YK, et al. 'Effect of Pain Reprocessing Therapy vs Placebo and Usual Care for Patients with Chronic Back Pain: a randomized clinical trial'. *JAMA Psychiatry* 2022;79(1):13–23.

neuropathic pain, which may be less sensitive to psychotherapy: Eccleston C, Hearn L, Williams ACdC. 'Psychological Therapies for the Management of Chronic Neuropathic Pain in Adults'. *Cochrane Database Syst Rev* 2015;2015(10):CD011259.

Conclusion. Become a Bowerbird

patellofemoral-pain syndrome. It's an overuse injury, often seen in runners and cyclists: Petersen W, et al. 'Patellofemoral Pain Syndrome'. *Knee Surg Sports Traumatol Arthrosc* 2014;22(10):2,264–74.

Teppo Järvinen's study, of course, featured prominently in the literature: Sihvonen R, et al. 'Arthroscopic Partial Meniscectomy Versus Sham Surgery for a Degenerative Meniscal Tear'. *N Engl J Med* 2013;369(26):2,515–24.

a recent review he co-authored of the challenges to developing effective analgesics: Jayakar S, et al. 'Developing Nociceptor-Selective Treatments for Acute and Chronic Pain'. *Sci Transl Med* 2021;13(619):eabj9837.

I must understand in order that I may believe: Clark K. *Civilisation: A Personal View*. London: BBC and John Murray, 1969: p. 44.